Super Seminars, Legendary Lectures, *and* Perfect Posters: The Science of Presenting Well

Ian Wilkinson, MSc, PhD, DClinChem, CSCC (Cert), FCACB, MBA

AACC Press
2101 L Street NW, Suite 202
Washington, DC 20037-1526
www.aacc.org

ISBN: 0-915274-94-9

Printed in the USA

About the Author

Ian Wilkinson, MSc, PhD, DClinChem, CSCC (Cert), FCACB, MBA, is the author of the best-selling *The Hitchhiker's Guide to Clinical Chemistry* and *Life, the Lab & Everything.*

Table of Contents

INTRODUCTION

Introduction

Master the art of presenting well!

Everyone would like to give super seminars, legendary lectures, and perfect posters. If you can't effectively present your ideas to an audience, nobody is going to pay much attention to you. If you can't address groups of colleagues, you're unlikely to be an effective manager. If you can't give presentations at meetings or workshops, you're less likely to make the connections that are essential to your career development. In short, if you can't effectively present your work, you may as well not have bothered doing the work in the first place.

Don't hide your light under a bushel. Present! Accept presentations as an essential part of your career development, just as you accept engaging in continuing education, taking on new responsibilities, or learning new skills. A presentation represents a wonderful opportunity to advertise yourself and your work before an audience that actually *wants* to hear what you have to say. Any opportunity to give a presentation is a chance to shine, to be noticed, and to make connections. Don't throw such chances away. You may be the most talented person in the cosmos, but others will never know it unless they hear you speak.

Sharing your data and ideas is only one small part of why you should be presenting. Presenting is also a means of getting the attention of people who may be important to your career. That is why it's vitally important to present well, in a polished and professional manner. Remember: You never get a second chance to make a first impression. One of the people in your audience may turn out to be your next employer or the person who recommends you for a job. One of the browsers at your poster may remember you several months later and invite you to give a paper. Presentation is a form of networking, and, in an uncertain job market, you never know when you may need your networking connections. The old saying is true: Success is a matter of whom you know. Networking is the secret ingredient that turns a job into a career.

What Business Are You In?

What business do you think you're in? Clinical chemistry? Hematology? Microbiology? Cytology? Whatever "ology" we use to describe our work, we are actually all in the same business: the information business. No matter what the area of specialization, all diagnostic services share the common element of information. In essence, the human body is a "black box." Much time, money, and energy are expended trying to find out what is going on inside the box, in extracting timely, relevant, precise, and accurate information about the patient's health status. Science is the art of extracting information and synthesizing it into knowledge.

Information has two key characteristics: its content and its means of transmission. The basic units of informational content are ideas, or "memes." Memes are the information world's equivalent to genes. Like genes, memes are bits of information passed on from individual to individual. And like genes, they love to reproduce.

To shift to a disease metaphor, memes also love to infect new minds. To do so, however, they must be transmitted effectively. They must be carried by vectors. Lectures, seminars, posters, books, newspapers, tapes, videos, movies, and the Internet are all vehicles for transmitting ideas; that is, they are vectors for memes. This book is about optimizing meme transmission by providing the right information to the right people in the right format at the right time. That's the secret behind the science of presenting well.

How This Book Is Organized

This book is divided into three sections. Part One, Communication, introduces key communication concepts. Part Two, Oral Presentations, provides tips on giving outstanding lectures, seminars, tutorials, and presentations. Part Three, Visual Aids, explains how to make your slides, overheads, and other visual materials work for you rather than against you. Part Four, Poster Presentations, presents techniques for producing and presenting a perfect poster.

Each part of the book begins with an abstract and ends with a summary of the material covered. This structure reflects the framework underlying every outstanding presentation:

1. Tell people what you are going to tell them.
2. Tell them.
3. Tell them what you told them.

Caveat Lector

This book is a guide to the science of presenting well, not the last word on the subject. Use your common sense as you apply the suggestions found within. Remember, there are few immutable laws. Most will bear some bending on occasion. Others can even be discarded altogether in certain situations. This book is a toolbox of ideas, not a list of commandments. It is a guide book, not a rule book.

I hope this book will help you make all your presentations outstanding!

Ian Wilkinson
Phobos, 1998

Part One: Communication

Part One: Communication

This section introduces essential communication concepts. Showing how our perceptions influence our judgments about people, it discusses the importance of body language and dress. It outlines the seven senses used by highly effective presenters: sight, hearing, touch, taste, smell, humor, and common sense. It also lays out the six major barriers to communication: physical, physiological, language, culture, education, and world view.

Perceptions

The First Four Minutes

People make judgments about you within the first four minutes of seeing you. They may never have seen your CV, read your autobiography, or heard what nice things your mother has to say about you. They may never have spoken to you, listened to you, or heard anything about you. None of that matters.

We all judge people within four minutes of meeting them, and we don't even know we're doing it. This is why first impressions count: You never get a second chance to make a first impression. No best of three, no timeouts, no retakes. You must get it right the first time.

Fashion Victims

Fashions come, and fashions go. In the world of presentations, however, fashion almost never changes. Remember, you are giving a presentation, not interviewing for a job as emcee on the Fashion Channel.

The trouble with fashion is that everyone has his or her own tastes. Things would be a lot easier if Chairman Mao had taken over as designer-in-chief at Armani. Then we could all dress in drab, genderless outfits. In the world of presentations, that fashion nightmare is not that far off. There is a uniform for success.

For men, the recommended uniform consists of the following items:

- A solid, dark blue suit. No patterns, no pinstripes.
- A plain white shirt. No patterns, no stripes.
- A dark blue tie. No stripes, crests, polka dots, tiny golfers, or Star Fleet insignia.

Why so boring? Because men's fashion is an oxymoron. As for interviews, wardrobes for presentations must be ultra-dull. Some men don't have to try very hard. Others do.

Remember: In the absence of a Maoist fashion manifesto, people have widely divergent opinions about fashion. You can upset people without ever opening your mouth, simply by wearing a loud jacket. When you are giving a presentation, you want all eyes to be on you, not on your outfit. You want the audience to pay attention to your message, not your boisterous pair of socks. Don't assault people's visual cortex with disturbing patterns, dots, or colors. Remember, 85 percent of a person's sensory intake is visual.

The recommended presentation uniform for women is equally dull:

- A solid, dark blue suit or a tailored skirt and jacket.
- A white, very pale yellow, or very pale blue blouse.
- A plain, dark colored scarf, if desired.

Given the truly fabulous array of body adornments available to women, these rules may seem like cruel and unusual punishment. Why should a woman dress as dully as a man? The logic is exactly the same as for men: You want to avoid distraction. You want the focus to be on your message, not on the way you look.

Be extra-careful about accessories. Because anything that shines or jangles will distract your audience, keep rings, bracelets, and lapel pins to a minimum. In lecture halls, where lighting tends to come from many different angles, even a large diamond ring can cause multiple reflections that shine into the audience's eyes.

The whole point of "dressing down" is to exploit your audience's sense of contrast. The contrast of the dark blue suit against the pale shirt or blouse creates a subconscious impression of precision and authority in audience members' minds. By dressing conservatively, you use visual stimuli to convince the audience of the validity of your work and yourself—an essential first step on the path to establishing credibility.

Body Language

The unspoken messages that create first impressions do not end with the clothes that adorn our bodies. Our bodies themselves are great communicators. Body language consists of the myriad actions and postures of our body and face with which we send unconscious messages to others.

For example, yawning while listening to someone is generally considered rude. So is looking around while someone is talking to you. Crossing your arms, putting your hands together, or clenching your fists signal defense or aggression. Standing with your hands on your hips indicates aggression or challenge. Opening your arms with your palms face up shows openness.

Similarly, avoiding eye contact or allowing your eyes to dart constantly signals uneasiness and induces mistrust in your audience—even if you're just shy. (Of course, in some cultures and subcultures, looking directly into someone's eyes has the opposite meaning. In the army, for example, a private does not look into the eyes of a superior officer.) Looking people in the eye is particularly important when you shake hands. There is nothing worse than contradictory body language, such as shaking hands but avoiding eye contact.

Almost all body language is subconscious. We are unaware that we are transmitting or receiving these messages, yet they are conveyed loud and clear. You might think that slouching on the edge of a desk or leaning on a lectern makes you seem cool or relaxed. The message you're conveying is actually negative, however. In fact, your body language detracts from your presentation's energy by slowing it down psychologically. Similarly, jiggling change in your pockets is impolite. Have yourself videotaped while giving a presentation. You'll improve your body language and identify irritating mannerisms you may be completely unaware of.

The Seven Senses of Highly Effective Presenters

Most people have seven senses: a sense of sight, a sense of hearing, a sense of touch, a sense of taste, a sense of smell, a sense of humor, and common sense. Most people manage to use five of these senses when they're communicating. Occasionally we use our sense of humor and common sense to enhance the information exchange.

Use these natural tools to communicate well. Capitalize on all of the senses by making your presentations as "senseual" as possible.

Sight

Sight is the most important communications tool. When people get together in physical space or on-screen, 85 percent of communication takes place by visible means. Your body language's nonverbal signals must not drown out or contradict the verbal message you are trying to get across. Don't give a verbal message about the importance of always being well-dressed while you're wearing cut-off jeans, grass-stained sneakers, and one of Bruce Willis' used T-shirts, for example.

Hearing

The second tool in our communications toolbox is our sense of hearing. In most people, this sense is underused and underdeveloped. Except for Vincent Van Gogh, we have twice as many ears as mouths. There may be a hidden message here. Always try to keep this 2:1 ratio in mind whenever you are trying to understand and to be understood.

Voices are often taken for granted as a communication tool, but the human voice can be a beautiful instrument: sonorous as a French horn, haunting as a flute, resonant as an oboe. By varying pitch, volume, and tempo, you can convey a wide range of emotions without ever uttering an intelligible word. You can turn a statement into a question, sharpen sarcasm's sting, and soothe a baby—or an entire audience—to sleep. The words we speak transmit only seven percent of emotional communication; our tone of voice transmits 22 percent.

Use your voice to add impact to your presentation. Listen to yourself on audiotape or, better still, on videotape. (Videotaping is better, because you can not only listen for voice problems but also check your facial expressions and gesticulations.) Do you talk through your nose? Do you let your voice drop off at the end of sentences? Do you mumble? Do you have a weak and screeching voice or a strong and resonant one? Do you enunciate clearly? Do you project your voice well?

Also beware of what I call "Urrmaarrhhspeak," a dreadful disease that sounds as bad as it is. It is endemic in politics, especially when politicians desperately want to emit sounds without actually saying anything. Listen to people being interviewed on radio and TV. Fearing the empty spaces between words, they insist on emitting long, drawling urrmmms and aarrhhs while thinking about what to say next. It's true that nature abhors a vacuum, but nature also abhors the vacuous. Tape yourself giving a presentation. See if you, too, suffer from chronic Urrmaarrhhspeak.

One of the best ways to truly master public speaking is to join the local Toastmasters Club, an organization dedicated to the art of public speaking. The Toastmasters are a gold mine for anyone interested in public speaking. You'll find information on local chapters in your phone book or at your local library.

Touch

The third tool in our communications toolbox is our sense of touch. At first, this sense may seem irrelevant. Unless you're at a Braille convention, what role does touch play in presenting well? A great deal, it turns out.

At the beginning and end of your presentation, you will have opportunities to shake hands with many people. Some of them may be very important to you, either now or in the future. Session chairs and symposium organizers are often eminent in their field, for example. They could be your next source of collaboration or even of employment. Remember: Sharing your data and ideas is only one reason to give presentations. Presentations are also a means of attracting the attention of those who may be important to your career. Presenting well can be an extension of networking.

Some audience members will want to meet with you after your talk, especially if they have unanswered questions. Always be ready with a stack of business cards and a well-honed sense of touch. Master the art of the handshake. An almost unnoticed, automatic ritual in our society, it is nevertheless vitally important. Handshakes should be firm but not vicelike. Limp handshakes convey a negative subliminal message about you. Hands should be dry, not dripping with perspiration.

Shaking hands can sometimes be difficult. Some people, for instance, stop in mid-air so that you end up shaking hands with their fingertips. To guarantee a perfect handshake every time, try these strategies:

1. Keep your fingers together.
2. Point your thumb straight out until it's almost 90 degrees to the side of your index finger.
3. Keep your thumb and your palm in the same plane.

4. Incline your hand toward your body about 45 degrees to the vertical.
5. Aim the area between your thumb and index finger at the equivalent area on the other person's hand. Use a swift and definite motion. Once the two areas connect, close your fingers around the other person's hand.
6. Look the other person directly in the eyes and smile to ensure a perfect handshake. There is nothing more disconcerting than shaking hands without making eye contact.

Smell

The fourth tool in our communications toolbox is our sense of smell. What on earth has smell got to do with presenting well, you ask? Olfactory matters are important—ignore them at your peril.

Halitosis, for instance, is evident from several feet away. At best, people in the front row will be disgusted; at worst, they will die of gas poisoning. Most halitosis-sufferers are not even aware they have a problem. Brushing one's teeth helps, but is usually not enough by itself. Using breathmints is another solution.

Smokers have several olfactory problems. Most smokers have an impaired sense of smell and aren't even aware that they smell like a stale ashtray. They do. They often have bad breath and stale-smelling clothes and hair, and the smell lingers long after they have returned from a quick smoke. Women who smoke often try to hide the smell with perfume. Unfortunately, this only compounds the problem.

In addition, smokers are becoming social pariahs. Most meetings and congresses are now smokefree. Nothing is more likely to ruin a wonderful first impression than a presenter who dashes out for a cigarette immediately after a presentation instead of mingling and networking with participants. All of this may sound unfair to smokers, but it is nevertheless true.

Humor

A sense of humor can be one of your most effective communication tools, as long as you use it with discretion. See Part Two for a discussion of humor.

Common Sense

The seventh tool in our communications toolbox is your common sense. Use this sense as often as is possible.

Extrasensory Perceptions

When did you and I begin to communicate? You might think it was when you finished reading the first sentence of this book. We actually began communicating long before that, at least as far back as the moment you first read this book's title or saw an ad for the book.

The same holds true for presentations. You may think your presentation begins just after you've been introduced to the audience or the first person reads the first sentence of your poster's introduction. You would be wrong. You begin communicating with your audience long before you see each other. If you're famous or at least known to your audience, the mention of your name marks the beginning of communication.

If you are not famous, communication begins with the title of your presentation or poster. That's why titles are so crucial. Like a newspaper headline, a title must catch an information-overloaded reader's eye and hold that person's attention. Avoid clichés, such as "a novel" whatever or "a unique" thingamajig. Try to make the title provocative, or at least enticing. "Cholesterol Testing Past, Present, and Future" is guaranteed to put even an insomniac to sleep.

"HDLs, LDLs, and Trigs: The Good, the Bad, and the Ugly" shows some imagination, but is a bit contrived. Phrasing a presentation title as a question or challenge can be an effective technique. "Cholesterol

Testing: Who Cares?" may be going too far, but something like "Does Cholesterol Screening Make Sense?" or "The Myth of Cholesterol Screening" is appropriate. You need to be provocative or challenging without being preposterous.

Having grabbed your audience's attention with your headline, you now need a few powerful first lines so that they will read the rest of your abstract or presentation description. Like the beginning of a good thriller, the first paragraph of your abstract should be enticing—an intellectual aperitif. Like a perfectly shaken martini, it should whet the appetite and get the intellectual juices flowing.

Compose your abstract with care. It is an advertisement you take out to promote yourself and your work. It is the very first thing that people will know about you, long before they have met you or read your resumé. Remember, you never get a second chance to make a first impression, even at a distance!

Barriers to Communication

Communication is essentially one brain's struggle to explain itself to others of its kind. The brain is trapped inside the body and has only a few primitive and highly restricted tools—the senses—with which it perceives and communicates with other brains.

There are six major barriers to effective communication: physical barriers, physiology, language, culture, education, and world view. If you fail to address these barriers, you could be a Ronald Reagan—known as "The Great Communicator"—and still be unable to get your point across.

Physical Barriers

The most obvious communication barriers—and the easiest to fix—are physical barriers. Nonetheless, presentation after presentation falls prey to them. One of the most common physical barriers during an oral presentation is that the audience, or part of it, can't see you, your slides, or your overheads. Or they can't hear you because there's collateral noise, your enunciation is poor, the microphone is broken, or you forgot to switch it on. Another golden oldie is the handwritten overhead. You might be able to read your meandering lines of scrawl, but no one else can.

Physiology

The second major communication barrier is physiological. To understand fully the importance of physiological barriers, we must turn not to physiology but to the so-called social sciences. Once upon a time there was a social scientist named Abraham Maslow. One day he made a list of the fundamental human needs and ranked them as shown in the accompanying illustration.

Illustration 1. Maslow's Hierachy of Needs

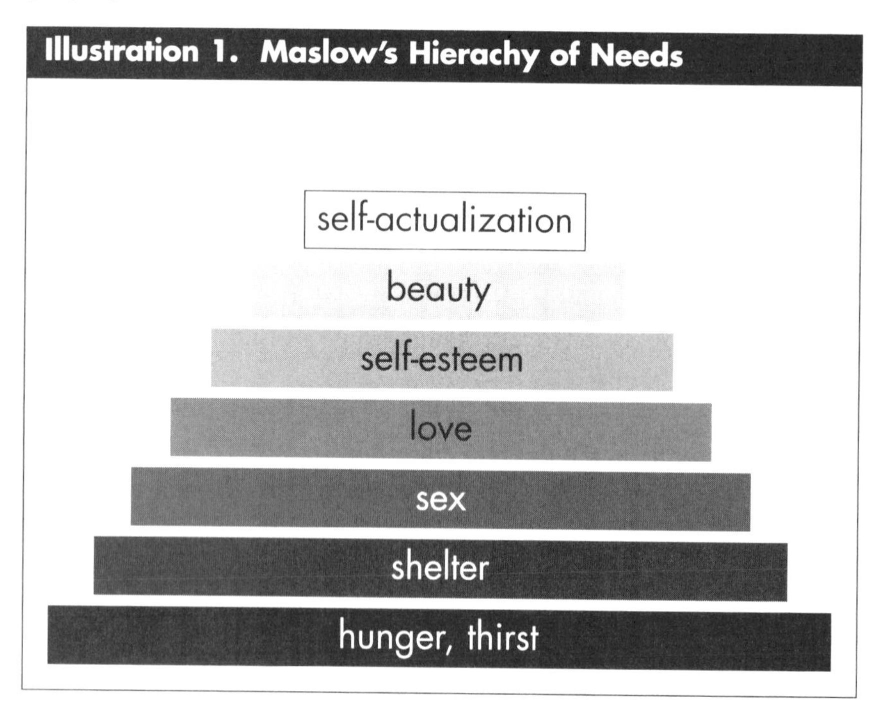

Keeping Maslow's hierarchy of needs in mind can help you avoid communication barriers. At the bottom of the hierarchy—the base of the pyramid—are the very basic, brain-stem type needs. According to Maslow, you cannot fulfill higher needs on the pyramid before you satisfy, at least in part, the lower needs. Physiological needs have a nasty habit of overriding people's intellectual needs, for example. If your audience is hungry or thirsty, too hot or too cold, or just plain tired, they won't be able to give their attention to you.

Try these tips for overcoming physiological barriers:

■ **Choose your time well.** To meet your audience's physiological needs, you could throw them peanuts, doughnuts, and coffee every few minutes, but a simpler way is to be careful when choosing your time slot. First thing in the morning is not a good choice, for example. People often arrive late, especially if it's the meeting's first day and they need to find a parking space, register, and find the correct lecture hall. Those who skipped breakfast may be hungry or in need of a caffeine fix.

Just before noon is a disastrous time. People have been concentrating all morning. They are beginning to tire and probably want to stretch, smoke, drink, or take their medication. They are almost certainly beginning to feel hungry, so their thoughts are more likely to be on soup and salad than upon correlation coefficients and intercepts. In addition, it is an almost universal rule that symposia run late. Unless there is an exceptionally skilled chairperson, most speakers go over their allotted time limit, and question periods compound the overruns. By the time the last speaker of the morning session gets up to speak, the session is often way off schedule and beginning to impinge on the audience's lunch break. Avoid this time spot at all costs.

What about immediately after lunch? If the lunch break was fairly short and not of the ethanol-based variety, you might consider this slot. People will be refreshed after having had the chance to eat, talk, walk, smoke, and burp for an hour. But try to avoid speaking during the second slot of the afternoon session. By then postprandial hypoglycemia has begun to set in. Some people experience an overwhelming urge to sleep one or two hours after lunch. If you can't get the first spot in the afternoon, go for the first spot after the afternoon coffee break.

If you're given a chance to choose any spot you want, choose the slot immediately after the morning coffee break. This is the least worst time.

■ **Arrive Early.** Arrive well head of your presentation. Ideally, you should visit the room your presentation will be in well ahead of the day itself. The most obvious benefits of this pre-presentation visit are that you know where the room is, how the furniture is arranged, how many people the room will hold, what kinds of audiovisual equipment is available, and where all the controls are. Of course, this advance work is impossible if you are presenting in Australia and you live in England. In some situations, such as seminars in your own department, you may have a great degree of control.

Check the room's temperature well in advance. Adjust thermostats and open or close windows or doors. Move the lectern to the side. Check visibility by projecting one of your slides or overheads and then sitting in various parts of the room. Make sure everything works: the microphone, the slide projector, and the room dividers.

If physiological parameters are within your control, plan for them, too. Arrange for coffee and zero-cholesterol, fully recyclable, high-fiber, low-UV muffins. Plan for regular breaks if you are giving a workshop, a long lecture, or two lectures back to back. Most people need breaks at least every 50 minutes. Kidneys are just about the only body part that can function without a break for hours on end. They don't need a break, but bladders do.

Language

Language is the third major communication barrier. Most likely you will be giving your presentation in English, since English has superseded Latin as the language universally spoken by educated people. English is one of the richest languages in the world. If you compare French, German, and English dictionaries, you'll see that the English dictionary is two or three times larger than the others. And there are many types of English: English English, American English, Australian English, New Zealand English, and myriad other forms. Don't feel you must write or speak the perfect English of Fowler.[1] Just write and speak as clearly and simply as possible.

Unfortunately, English is rapidly being replaced by Jargonese. Jargonese is particularly popular with molecular biologists, computer programmers, and pop psychologists who appear on chat shows. Avoid jargon whenever possible. When you must use jargon, be sure to define it. This is especially important for poster presentations. Keep in mind that 95 percent of the people who will view your poster will not be specialists in your area. Excluding these people by using undefined jargon is impolite and not too sensible. Also avoid acronyms that will be understood only by other experts in your field. Remember, the words used in science are precision-crafted tools. If you are to be successful in bringing comprehension from chaos, you must use words with precise, universally accepted meanings.

Culture

Culture is the fourth major communication barrier. What does culture have to do with communication? The next time you are at a lecture, seminar, or poster session, look around and see how many races, ethnic groups, and nationalities the audience represents. Your message will be interpreted though the lens of each person's culture. Developing an awareness of cultural differences and adapting your presentation accordingly will help you avoid antagonizing your audience.

For example, the roles of women, old people, and young people vary greatly across cultures. Some cultures revere their elders; others mug them. Some cultures believe that women have no place in science, in

[1] Books on writing range from the impenetrably dull, such as *Fowler's Modern English Usage*, to the sublime, exemplified by *The Economist Style Guide*.

meetings, or in public—and certainly not in short skirts. Some societies believe that consensus and the common good are more important than the rights of the individual.

If you will be visiting a different culture, do your homework before you arrive. Once you get there, observe and listen. Staying culturally aware is harder to do when you are hosting people from other cultures or addressing subcultures within your own society. You could argue that it's your country and this is how we do things here. Perhaps. But try to be aware and avoid offense where reasonably possible. It's easy to offend people without even knowing it. The Japanese, for example, have a special way of presenting and receiving business cards: It is impolite to simply glance at a card and casually tuck it into your wallet. The difference is subtle, but the potential for offense immense. To succeed in our increasingly multicultural world, you must "internationalize" your mind and become culturally aware.

Education

The fifth communication barrier is education. Your audience's educational attainments may vary widely in breadth and depth. That makes it difficult to judge what level to pitch your presentation to and what assumptions you can safely make. Few people have a perfect pitch when singing, throwing a baseball, or giving a presentation. If you see people cleaning out their wallets or practicing their golf swings, you are pitching your presentation too low. If you see people with glazed looks or puzzled expressions, you have either pitched your presentation too high or your audience members were all at the same party you were last night and the aspirin has yet to take effect.

To achieve perfect pitch, keep these suggestions in mind:

■ **Know thy audience.** Knowing your audience is the key to success. That's usually a straightforward task for lectures and seminars given to students. Find out who they are, what their background is, what other courses they have taken, and what content they expect to learn from your presentation. Then pitch your presentation to the lowest common denominator.

Finding the lowest common denominator for presentations at symposia is much more difficult. Obviously, a meeting entitled "The VIIth International Symposium on Those Small Green Things with the Raffia Base" would suggest an audience that is highly specialized and highly knowledgeable within the realm of small green things with a raffia base. A meeting entitled "The XIXth National Congress on Wacko Mindboggling Science Stuff" would suggest a much lower common denominator.

To estimate the lowest common denominator for workshops, include a brief background questionnaire in the workshop registration form. For poster presentations, design your poster for the 95 percent of the audience who are not specialists in your field. Minimize jargon, resist the use of acronyms and abbreviations, and explain everything fully—even the things that seem terminally obvious.

■ **Know thy stuff.** Knowledge is a double-edged sword. I have assumed that as a presenter you know your stuff. Unfortunately, that's not always the case. Sometimes people are forced, coerced, blackmailed, or tricked into presenting against their better judgment. Supervisors, department heads, or other boss types may try to "guilt" you into presenting. Although giving presentations is important for your career and personal development, there are legitimate reasons for saying no. If you don't know a topic very well, do not attempt to make a presentation on it.

If you do know your stuff, you'll find that giving a presentation will help you really get to know your subject inside and out. Preparing your presentation forces you to clarify key concepts. If you give an annual lecture on thyroid disease, for example, you'll find that revising your notes and overheads each year helps you make your points more clearly and simply. It's as if we were sculptors: Each time we look into a subject, we chip away a few more pieces of extraneous material to reveal the masterpiece hidden within the rock.

There's a strong positive feedback loop in operation here. To be an outstanding speaker, you must know your subject inside and out. The best way to know your subject inside and out is to give presentations on it.

World View

The last major communication barrier is your world view. Your world view represents your interpretation of all the experiences and knowledge you have accumulated to date. Depending on their world views, two people can look at the same object and see two different things. Although these perceptions are apparently incongruent, both are equally valid and both are equally true.

It's very difficult to see the world from someone else's perspective. In fact, many of us find it hard to understand our spouses, our children, or other relatives. It's even more difficult to understand the world view of a person from a different culture. Despite these difficulties, it's important to remember that radically different world views exist. Keep that in mind when making assumptions, choosing language, and using metaphors in your presentations.

Retention

Presenters want their audiences to remember their presentations—because they were outstanding, not because they were disastrous.

Making presentations memorable doesn't just happen. You must make it happen. A good presenter has prepared, practiced, and applied the principles outlined in this book. There's no point in sharing your message if people don't remember what it was the moment they leave the lecture hall. Understanding how memory works is the first step in becoming memorable. Researchers know that we remember only 10 percent of what we read, for example. Does that mean you must read this book at least 10 times if you want to remember it all? In reality, memory is not that simple. We focus our attention on some things and have little difficulty remembering them, but we can have immense difficulty remembering other things that are no more complex.

We are much more efficient at listening. We remember 20 percent of what we hear, nearly double the retention rate for reading. This marked increase in the efficiency of learning may come from listening more actively to others than we do to the voice that reads along with us in our minds. Even listening to a book on tape is more memory-efficient than reading the same book in printed form. This doubled retention rate is one good reason to prefer giving an oral presentation to a poster presentation. In fact, an oral presentation that includes visual aids has an even more profound effect on retention rates. That's because we remember 30 percent of what we see. If you want to give a memorable presentation, carefully prepared slides or overheads are vital.

How can you boost retention rates even higher? The answer is obvious: Combine sight with hearing. We remember 50 percent of what we see and hear. That's why TV ads are seldom silent. In fact, the volume often goes up during commercials. Advertisers want to bombard our minds with sound and vision, knowing that each reinforces the other.

Fifty percent retention sounds pretty good. By using sight and sound we can ensure that about half of our message will be remembered. Of course, that means that half will still be forgotten. With passive, one-way learning situations, such as oral presentations, that's about the best you can do.

There is a way to get retention rates even higher, however: audience participation. Getting people to give a presentation increases their personal retention rate to 80 percent. Getting a person to actually perform something for themselves increases retention to 90 percent. Let's say you want to learn how to operate the Pangalactic Hyperwhiz Omnianalyzer. If you read the manual, you'll probably remember about 10 percent of the information. If you listen to a teleconference on how to operate the machine, you'll retain 20 percent. If you attend a presentation replete with slides showing various steps in Omnianalyzer operation, you'll retain 50 percent. If you give a presentation to colleagues interested in learning more about this amazing machine, you'll retain 80 percent. If you start using the Pangalactic Hyperwhiz Omnianalyzer—pressing every button, checking out every computer screen, and performing analytical runs—your retention rate will soar to 90 percent. We learn best by doing. If you want to really remember this book, give a presentation on how to give outstanding presentations. When you have to do something yourself, you have to understand it. In understanding it, you remember it.

Summary

Part One introduced the essentials of communications. Grounded in a discussion of how our perceptions influence our judgments about people, the section began by explaining why proper dress is essential during presentations and how body language can enhance or undermine your message. The section then outlined the seven senses of highly effective presenters: sight, hearing, touch, taste, smell, sense of humor, and common sense. The section then introduced six barriers to communication—physical barriers, physiology, language, culture, education, and world view—and offered strategies for overcoming them. The section concluded with a discussion of how to help audiences remember what they have learned. In the next section, we'll put these new ideas into action as we learn more about oral presentations.

■

Part Two: Oral Presentations

Part Two: Oral Presentations

Expanding upon themes introduced in Part One, this section applies the strategies suggested there to oral presentations such as lectures, seminars, or business presentations. The section begins by suggesting strategies for reducing stage fright and building credibility. The section then lays out the anatomy of a presentation, from start to finish. Separate sections give hints on language and equipment.

Stage Fright

If you're like most people, you look forward to giving oral presentations about as much as you look forward to having a hole drilled into your head with a blunt, rusty spoon. That's understandable but unfortunate. An oral presentation is a wonderful opportunity for advertising yourself and your work before your peers. Do not throw away this chance to shine. You may be the most talented person in the cosmos, but people will never know it unless they hear you speak.

"OK," you say. "But if oral presentations are such wonderful opportunities, how come so many people feel ill at the very thought of them?" The answer is simple: fear! Fear of public speaking is number one on the list of phobias. (Death is way down the list, coming in at number seven.) Most people would rather be thrown into a bathtub full of snakes or spiders than give an oral presentation.

Overcoming your fear of public speaking should be one of your highest priorities. You may never reach the point of being perfectly fearless when it comes to presenting. That's OK. While too much adrenaline brings out the flight response rather than the fight, having just a little adrenaline flowing can increase your concentration and enhance your performance.[2]

Remember, great orators can be made as well as born. You *can* overcome your fears. In my own case, I went from being so shy I couldn't even talk to myself to being able to give award-winning lectures to classes of 150 or more students. There are various approaches to overcoming fear, including avoidance, flooding, cognitive psychology, and practice.

Avoidance

There is one guaranteed, 100 percent effective way of completely ridding yourself of fear: Never give an oral presentation.

This is not a sensible option. Addressing groups of co-workers helps you be a more effective manager, for example. Giving presentations helps ensure that people pay attention to your ideas. And giving presentations at meetings helps you make essential career connections. Every time you present, you advertise your work, your ideas, and yourself. Overcoming your fear and accepting opportunities to present are just another part of your career development.

Flooding

Not to be confused with what happens when you live too close to the Mississippi, flooding is the fine art of scaring yourself to death. How does that help you overcome stage fright? The idea is simple. Imagine everything that could possibly go wrong with your presentation. What if you show up late or on the wrong day, for example? What if you forget to shave or put on your makeup? What if you give the entire presentation with your fly undone or your slip showing? Or your pet iguana eats your overheads the night before your talk? Or the audience is so hostile everyone has to be muzzled with Hannibal Lecter masks?

---

[2] *Anxiety and Panic Attacks: Their Cause and Cure*, by Robert Handly and Pauline Neff, describes a five-step program the authors claim will help you conquer your fear, reduce or eliminate anxiety, end stress-related illness, change your bad habits, and more.

Once you have thought of all these terrible things, start over and think of things 10 times worse, and then 100 times worse. Do this over and over. Showing absolutely no self control, wallow in your fear. Indulge yourself in sheer terror. Overwhelm yourself with your worst nightmares. Let your heart pound, your pulse rise, and your adrenaline flow by the bucketful. Scream and cry if you feel like it. Keep this up as long as you can.

Fortunately, there's a limit to how much you can scare yourself. After a while, you'll get tired of being scared out of your wits. No matter how hard you try, you won't be able to scare yourself anymore. You have reached a saturation point. Slowly your pulse rate will drop, and your breathing will return to normal. Rest for a while, thinking fun thoughts about a date with Keanu Reeves or Michelle Pfeiffer dressed as Catwoman.

Once you've calmed down, scare yourself all over again. As before, dream up graphic details about everything that could possibly go wrong. Once again you will reach a plateau of fear you cannot go beyond. Again your body will slowly return to normal, and your fears will subside. Rest for a while, think of something fun, then do the whole scare yourself to death routine all over again.

After several cycles of fear and relaxation, you'll notice that the degree of fear decreases each time you shock yourself. Your ability to scare yourself diminishes with each use of the images. You are becoming tolerant to your fear. By using this technique, you essentially flood your fear response so that it can no longer respond. This technique can be very effective for extremely nervous persons, especially those about to make their first presentations ever. The downside of this approach is its tendency to be tiring and psychologically draining. Of course, there's also the risk of cardiac arrest.

Cognitive Psychology

Cognitive psychology sounds complex, but the idea is simple: You feel the way you do because of the way you think. Instead of letting your fears envelope you, you face them head-on. Examining each fearful feeling, you defeat it through logical reasoning.

Many people have a hard time accepting this. They insist that it is their boss or their spouse who makes them feel bad. But nobody and nothing can make you feel bad. You can only do that to yourself. It is your interpretation of words or events that causes you to feel bad or good, not what somebody said or did.

As a result, cognitive therapy offers one of the most powerful means of recovery available for anyone who has ever felt lonely, depressed, insecure, anxious, worried—or terrified about giving a presentation. This isn't pop psychology. Cognitive therapy is a scientifically tested and proven approach to helping people overcome anxiety and fear and achieve personal happiness and growth.

Our problem is that we think too fast. The moment we start imagining a situation like giving an oral presentation, we start making assumptions. These subconscious value judgments are so fast, so ingrained, and so automatic that we aren't even aware of them. Our brains generate feelings the same way they generate the hundreds of instructions needed to tell our legs to walk. You don't think about it; you just do it.

Cognitive psychology works by intercepting *input* to our brains before the brain can translate it into a *feeling*. David D. Burns, MD,[3] one of cognitive psychology's most famous proponents, has identified common cognitive distortions and suggested techniques for overcoming them. Try one.

Draw a line down the center of a piece of paper. In the lefthand column, jot down your automatic thoughts about public speaking. On the right-hand side, logically refute those thoughts. One of your automatic thoughts might be, "My presentation will be a disaster." OK, what specifically will make it a disaster? "I'll forget what I am going to say." So what? "I'll look foolish." Why will you look foolish? Has anyone else in recorded history ever lost their place during a presentation? "Well, yes, I guess so...but that doesn't matter because I know that everyone will think I'm useless." How do you know that's what they'll think? Will everyone in the

[3] His *Feeling Good: The New Mood Therapy* and *The Feeling Good Handbook* offer clear explanations of cognitive therapy plus real-life examples and exercises.

audience, including your friends, think this? And even if one or two people do think this, so what? "Well, it will mean I'm inadequate." Why? Are these few people the universal arbiters of truth? Are they holding your child hostage and threatening to kill him if you make a less than perfect presentation? And so on.

Practice

Practice! Practice! Practice! "Practice makes perfect" may be a cliché, but it's true that the more you present in public the easier it becomes. Practice enables you to use the rational parts of your brain to overcome deepseated, automatic responses to unknown fearful situations. Thanks to the fight or flight response, our natural response is to flee from danger—and avoid public speaking. By practicing, you can learn to control your fears and use adrenaline to help you perform even better.

Practice begins long before you ever see your audience. At home or in your office, go through your slides, overheads, or notes and practice your presentation over and over. (Of course, this means you have to have slides, overheads, and other materials prepared well ahead of time. Never leave this to the last minute!)[4] By making your practice sessions as realistic as possible, you can make sure your presentation isn't too long or too short. You can make sure your slides or overheads are legible and loaded correctly. You can master the simple mechanics of presenting, such as flipping overheads, advancing slides, and using a pointer. Ideally you should have at least one dress rehearsal in the venue itself to familiarize yourself with the lighting switches, dimmers, microphone, pointing devices, slide projector controls, and so on.

[4] Advance preparation can help you avoid a common error: projecting slides every which way but up. To prevent disasters, place a large blue dot on the top right-hand corner of each correctly oriented slide and number the slides in sequence. (Don't write directly onto the slides, since you may wish to change the sequence in subsequent presentations.) The first time you drop the slide carousel just before presentation, you'll be happy you took the time. Use a similar technique for overheads. Place each overhead in a cardboard frame and stick a self-adhesive paper dot on the top right-hand corner of each correctly oriented overhead. Then number the overheads in sequence.

In addition to these actual practices, you should also practice in your head. Play through in your mind what you intend to do in reality. Dennis Waitley, author of *The Psychology of Winning*,[5] is the best-known advocate of this approach. He has trained Olympic athletes how to imagine every moment of their performance in their minds before they ever step into the arena. By building clear mental images of success in their minds, the athletes were able to excel in the real world of the sports arena.

You can use the same techniques as you prepare for public speaking. Your subconscious mind cannot tell the difference between what you *imagine* and what is *real*. If you constantly feed your subconscious images of success, it will accept them as fact. Chances are you'll give a better presentation once you actually step before an audience. Conversely, feeding your subconscious a constant stream of negative thoughts can create a self-fulfilling prophecy of goofups, embarrassment, and failure.

Once you have practiced a few times on your own, ask a friend or colleague to watch and give you feedback on your style, your tone of voice, and your presentation's overall clarity and logic. Better yet, watch a video of yourself practicing or actually presenting a presentation. Many facilities now videotape speakers as a matter of course, so you may be able to watch past performances for ideas on how to improve future presentations. Failing that, ask a colleague to videotape you.

There's nothing quite so educational as watching yourself on videotape. You can spot every fidget, every "urrmm," every gesticulation, every eye movement, and every smile or frown. There is no better way of judging our own body language, eye contact, movement, and intonation. In fact, videotapes are a far more effective source of feedback than friends or colleagues because you can step outside yourself and watch your performance with a critical eye. By identifying problems and working to correct them, you will face your next presentation with far less trepidation. [6]

---

[5] You can order his audiotape set from Larimi Communications, Dept. 11, 246 West 38th Street, New York, NY 10018, 212-819-9300.

Credibility

To be an effective communicator, you must be convincing. To be convincing, you must have confidence. To be confident, you must know your subject, prepare carefully, and speak publicly as often as possible. Being well-prepared not only enhances your ability to transmit a message to your audience, it also allows you to transmit a subtle, more subliminal message: that you are a well-organized, clear-thinking, competent person.

Preparation begins long before you're even invited to give a presentation. In fact, one of the most important steps as you prepare for a presentation is to establish credibility. The audience must *want* to listen to you. They must believe that you have something worth listening to, something that will transform them into active learners rather than passive couch potatoes waiting for a sea of information to wash over them. Credibility makes people stop doodling or biting their nails and sit up and listen to what you have to say.

How can you establish credibility if your name doesn't happen to be Einstein or you haven't won a Nobel Prize lately? Unfortunately, credibility cannot be bought at a supermarket or mail-ordered to arrive in a plain brown wrapper. There are several ways of establishing credibility: qualifications, expertise, believability, and likability.

Qualifications

We live in a world increasingly obsessed by framed pieces of paper that hang on people's office walls and attest to how smart they are. There are

[6] Practice isn't the only way to improve your presentation techniques. You can also watch others. Whenever you are in the audience of a lecture, seminar, or other presentation, make a point of noticing the presentation's good and bad points instead of letting yourself drift off into space. How clearly did the speaker transmit key concepts? Did she budget her time correctly? Were her slides or overheads well-designed? Did she handle questions well? Did she provide well-designed handouts? Did she cower behind the lectern or pace around the stage like a caged animal? What was her take-home message? Make your attendance at any presentation a double learning experience: Think not only about the presentation's content but about the way the information was presented.

"graduates" in everything from medical laboratory technology to certified canine esthetics, from clinical chemistry to toenail technology. We live in a certified world.

In many ways, that's a good thing. Certification raises standards and ensures that professionals are well-trained. And certificates are widely recognized measures of achievement attesting to a base of knowledge or set of skills. Whenever possible, obtain certificates of your own to attest to your skills. Be sure to always have a brief, up-to-date biography outlining your qualifications available for use in conferences' promotional materials. Always write your own biography.

Expertise

At one time, you had to have actually done something—such as spent years researching the mating habits of frozen fish—before you were considered an expert. That's no longer the case. We live in a world of instant experts.

In the world of science and medicine, however, expertise still has to be earned. In our field, work experience, continuing education, teaching, research, writing, and managing all count. And although you may not know it, we all have expertise to offer. You may be an expert at troubleshooting the Pangalactic Hyperwhiz Omnianalyzer or performing phlebotomies or working with computers. In my case, it's presenting well. Whatever the subject, your expertise adds to your credibility.

Believability

Why do we believe people like Walter Cronkite and distrust others like Richard Nixon? Often it is because of history. We may know that one person stands for truth and another for terminological inexactitudes. Believability is largely a function of perception—of your perceived value. Believability is a multifaceted beast: Qualifications, work history, and affiliations may all play a role, as do intangibles such as appearance and body language. In fact, these intangibles can be as important as your qualifications.

Great speakers are like actors or preachers, selling their message with passion and a natural flow of words. They convince because they are convincing. You wouldn't want to watch a movie in which the actors read their lines from a book. You wouldn't stand (or rather sit) for it, and neither should *your* audience. There is no more certain way of killing a presentation than by reading it. The only time you should read anything is if you are quoting from a text.

Get excited about your topic! Enthusiasm is infectious. Unfortunately, many presenters stand up and drone on and on monotonously about how exciting their topic is. This creates cognitive dissonance: The tone and body language do not match the message. Present energetically and enthusiastically, not as if you were under a death sentence. No salesperson ever sells anything without enthusiasm for his or her product. You, too, are selling a product: your lecture, your topic, your work, and, most importantly, yourself! Your enthusiasm will help your audience believe in you.

Likability

Love me, love my talk! We are more likely to listen to people we like. There are thousands of ways to turn people off, some of which require very little effort. Your audience could dislike your outfit, hairstyle, accent, tone, or laugh, for instance. They may dislike your seriousness or your sense of humor. They may just dislike you.

Fortunately, there are ways of establishing rapport and chemistry with your audience. The easiest is a smile. It is amazing how many people get up to speak looking like they have just been forced to listen to the complete works of Barry Manilow. Although chimpanzees grin in response to fear, in humans a smile is a universal sign of openness. Smile and others will smile back, with their minds if not their faces. Smiling can also make you feel more relaxed, which is always handy in a potentially tense situation.

Try beginning your presentation with a funny icebreaker. A little humor can be a big help in refocusing the audience's attention. Use humor carefully and sparingly, of course. While a tasteful joke can establish likability, a joke in poor taste will destroy your likability. Avoid racist and sexist jokes at all cost. And beware of self-deprecating jokes. Rodney Dangerfield can get away with not getting any respect. You cannot. After all, you are trying to establish credibility. Joking about what a klutz you are may get you a brief laugh but will ultimately undermine your efforts.

Being gracious will also create likability. If this is your first time in Armpit, Manitoba, say so. Tell your audience how wonderful it is to be here in Armpit, home of the world's largest mechanical mosquito. Thank your audience for inviting you. Glance at a local newspaper or TV news broadcast so you can work in a comment or joke with local interest at the beginning of your talk. Because you've established a link with them, the audience will see you as less of an outsider. Use your common sense, however. If you really are presenting in Armpit, Manitoba, do not tell jokes about small towns and baseball-cap-afflicted guys who drive pickup trucks.

With small audiences, try introducing yourself to people as they arrive. Shake their hands, exchange small talk, and remember to smile. And give them your business card. If they like your talk, they may want to talk with you about it sometime in the future or they may think of you when they're looking for speakers for their next convention. They'll probably give you business cards or at least scraps of paper with scribbled contact information in return, which helps you build your network.

Cultivate the art of remembering people's names. This is not only polite but also makes people feel good. People like to be known. One way of ensuring that you remember a new acquaintance's name is to immediately repeat the person's name three times in your mind while looking at him or her. Another is to immediately write down their name and read it over several times while imagining their image. (Exchanging business cards will help you put names in your long-term memory.) Or try associative techniques or visual mnemonics. You might focus on certain features to help you remember people's names, such as bald Brian or sandy-haired Susan.

Put those names to use to establish even more rapport. Work the name of someone you just met into your talk, saying "Ms. Scully was just telling me how important the hunt for Elvis has been to the X-Files project." Again, this helps change you from an outsider to an insider in the minds of the audience.

If you're giving a series of lectures to a small group of people, consider setting up a cookie club. Each week you take turns bringing in cookies (or celery for the calorie conscious) to eat during the break. Food is a universal icebreaker. Try it.

Anatomy of a Presentation

Every presentation has a beginning, a middle, and an end. Here are some suggestions for each step along the way.

Context

If your presentation is part of a symposium or lecture series, start intelligence-gathering as early as possible and certainly before you start preparing slides or overheads. Find out what topics are being covered in presentations before and after yours. If possible, ask other speakers for brief synopses of what they plan to cover. Doing so will help you gauge what level of complexity to use and help you avoid duplication. Don't panic if a presenter earlier in the day surprises you by covering an area you had planned to cover. Instead of going over the material again, use your allotted presentation time more effectively by skipping the background that person has already covered and getting straight to the meat of your presentation.

Arrive Early

When blocking out time for your presentation in your appointment book, don't just note the time that you'll actually be talking. Remember to allow time both before and after your presentation. Before your talk, you'll want to make any final preparations, renew old acquaintances, and make new ones. Afterwards you'll want to meet members of your audience.

Repetition

Repetition is an important weapon in the armory of the effective presenter. Repetition increases the likelihood that your audience's brains will absorb your ideas. Use this three-phase approach to repetition for all of your presentations:

1) Tell the audience what you are going to tell them.
2) Tell them.
3) Tell them what you told them.

You should also use visual repetition. Be consistent when choosing typefaces, styles, colors, and other elements throughout the presentation. You could use your organization's logo or another icon to enhance this feeling of unity of design. Put the icon in the lower righthand corner of every slide or overhead, where it will be discreet yet noticeable. Remember, repetition is important. Remember, repetition is important. Remember, repetition is important.

Word Budget

Like finances, presentations work best when you've made a careful budget. A word budget ensures that your presentation is balanced. It helps you guarantee that the amount of time you spend on each part of your presentation is proportional to that part's importance. Spinning out a lengthy anecdote during your introduction may help you relax and get your audience giggling uncontrollably, for example, but spending so much time on it may leave you pressed for time when it comes to getting your message across. A budget can also ensure that your presentation doesn't run over its allotted time or eat into the question-and-answer period.

In the sections that follow, you'll find some general guidelines that can help you achieve a balanced word budget. In the examples below, you'll find suggestions designed for the popular conference format of a 10-minute presentation. If your presentation will be longer, simply multiply the number of words per section but keep the ratios between sections the same.

In the Beginning

Too many presenters leap into their presentations without explaining why their topics are important or why their audiences should care. They don't state their topics or their theses. They just leave it up to their audiences to figure out what their presentations were about and what the take-home message was.

Don't be like them. View your presentation as a story you want to share with your audience. Like any good story, your presentation should have a beginning, middle, and end. Its beginning should contain your introduction, topic sentence, and thesis sentence. For a 10-minute presentation, spend no more than 150 words from your word budget on this part.

Use your introduction to set the scene. Help your audience figure out how your presentation fits into the big picture. Using a mental map can help them visualize how your presentation's logic will flow. (See Part Three for more information about mental maps.) Remember that people sum you up within four minutes of encountering you. Don't blow it! After your introduction, give your topic sentence. This single sentence should state exactly what your talk is about. "Why do I need a topic sentence?" you might be thinking. "Surely the title of my talk says it all." Perhaps. But your presentation's title probably doesn't say exactly what your talk is about.

In an effort to garner attention and attract interest, for example, you may have entitled your talk "The Endocrinic Effects of the Leather Goddesses of Phobos." That's attention-getting, but it doesn't tell your audience enough. Your topic sentence might be, "This presentation will describe our research on the effects of interactive computer games on the psychoendocrinology of adolescent males." ("Leather Goddesses of Phobos" was a popular computer game of the 1980s.) Your topic sentence must encapsulate your presentation's essence. It should be clear, concise, and so self-explanatory that the vast majority of your audience will understand it easily.

If you're afraid you'll flub your topic sentence, write it on a cue card. Inspiring self-confidence, the cue card will help you set off on the right foot. Of course, this should be the only thing resembling speaker's notes you have. You know your presentation inside and out, remember?

As its name suggests, the thesis sentence is a clear, concise sentence stating your presentation's thesis. "But I don't need a thesis sentence," you protest. "I'm going to talk about the disaster recovery plan that will help our lab recover from the emergency. It's not research, so I don't have a thesis!" Wrong. You *do* have a thesis. Whether you're presenting research or simply making a case, your thesis is the take-home message, the bottom line, the sound-bite.

For example, a presentation called "Stealth Fighters: Making Magic Bullets Invisible" might have a topic sentence like this: "In this presentation, I will describe how we can modify magic bullets to make them invisible to the body's defense systems." Here's a potential thesis sentence: "Modifying magic bullets with polyethylene glycol makes them more effective." When you boil my explanations, my tables, and my graphs down, that's what you get.

If you're having trouble coming up with a topic sentence or thesis sentence, you might need to focus your presentation a bit more until you find a unified theme. Have you tried to introduce too many new or unrelated ideas in one presentation? Or maybe you're trying to tack a thesis sentence onto an existing presentation. A thesis sentence shouldn't be an afterthought! If it is, you're designing your presentation in reverse.

To design a presentation the right way, take a piece of paper and write the single message you want your audience to take home with them.

Everything else in your presentation should flow from that one statement. List key points to support your thesis, support your key points with tables or graphs showing your data, and then come up with a title and topic sentence. See how your whole presentation hangs on the framework of that one sentence? By designing your presentation this way, you'll be guaranteed a cohesive, logical presentation. With your thesis sentence in mind, your audience will have no trouble following you as you flow smoothly from key point to key point.

In addition to enhancing your audience's understanding, your thesis sentence also sets the tone. It must transmit your enthusiasm, commitment, and conviction about the topic. You can transmit some of these qualities through your inflections, rhythm, and tempo. Your choice of words is important, too. Say "I believe..." rather than "It could be that..." Make sure how you're saying something matches what you're saying. You can't expect your audience to be enthusiastic about your topic if your speech and body language aren't.

First Impressions

Action movie directors know how to make an attention-capturing first impression: Begin a film with a high-speed car chase. Editors and journalists know, too. Newspapers grab your attention with glaring headlines, such as "Elvis Elected First President of Australia," and then offer tantalizing details in the first paragraphs of a story. And which opening do you think a novelist is likely to choose:

> *I pulled the pin from the grenade, shot the leopard in midair, pulled the poisoned arrow from my thigh, and fought desperately to land the Stealth bomber on the erupting volcano."*

Or:

> *"I woke up on Monday morning at 23 minutes to seven. Actually it was 24 minutes to seven. My clock was fast. I looked at my clock and thought, "My clock is fast."*

You get the picture. The same principles apply to presentations. You must immediately grab and hold your audience's interest and attention. If you don't capture their interest within the first four minutes, you have lost them for good.

In the Middle

The middle part of your presentation is where you provide background and key points to support your thesis. For a 10-minute presentation, you should devote about 150 words from your word budget for background and anywhere from 100 to 150 words for each key point. Remember, these word budgets should increase proportionately for longer presentations.

The background section is where you give your presentation's who, what, why, when, where, and how. Second-rate presenters tend to concentrate on the who and how of their work, but fail to cover any of the other bases. That leaves the audience unsure of the presentation's relevance. Spell it out for them: Briefly describe who did the work, what they did, when they did it, and where. Most importantly, explain how and why they did the work.

Key Points

The bulk of your presentation's middle section should be given over to a series of key points supported by data or examples. But don't get carried away. Never include more than seven key points in a single presentation. Despite our highly evolved brains, we have a hard time assimilating new information in a short period of time. On a good day, most people can assimilate a maximum of seven key points. Under the stressful conditions of marathon sessions at a conference or meeting, that number may be considerably reduced. Tired, hungry, or bored audiences have difficulty concentrating on anything, never mind assimilating new information.

Introduce your key points in the *reverse* order of importance. That way each acts as a stepping stone to the next. Your last key point should be the most important. Be sure to restate your thesis sentence with each key point, including the last. Remember, repetition is important.

How do you go about developing your key points? Sometimes it's not immediately obvious what your key points are. Let's say the International Macintosh Society has asked you to give a presentation on your outstanding research on rainwear. You have 30 minutes to fill, and hundreds of tables and graphs of data comparing various forms of rubber rainwear to vinyl.

As we discussed above, step one is to decide on your specific topic area. You decide that your latest work on the advantages of vinyl Macintoshes will be of most interest to this particular audience. Your thesis sentence becomes, "Extensive field trials have shown that vinyl Macintoshes are better suited to English summer weather than are rubber ones." Only now should you start thinking of key points.

You have a veritable flood of data from these studies. Which are important? The best way to decide is to brainstorm (wearing your mac and Wellington boots, of course). Freeing your mind of prejudice or judgment, jot down any idea that comes along. This is hard to do. We almost always make a split-second judgment about every idea we have. However, the key to successful brainstorming is to generate and record all ideas without judgment. Once they are down on paper, then you can consider each of them and reject those that don't support your thesis sentence. Brainstorming is most productive if several people are involved, especially if those people are the team that has worked on a project.

Once you have weeded out the losers and generated your list of key points, rank them from least important to most. Now develop them. For each key point, ask yourself the following questions:

- What data or examples do I need to support this point?
- What does the audience need to know about the methods used to generate this data?
- How can I show the data clearly and succinctly?
- What additional information can I provide to support and clarify this point?
- Is there a metaphor, analogy, or allusion I could use to help communicate this idea?
- What slides or overheads do I need to clearly and concisely support this key point?

To develop a key point, you must clarify and simplify. You might be tempted to write your key point out. Remember: A key point is a single, succinct point, not a paragraph or a novel. How many times have you been to presentations where the presenter put up endless overheads covered in an illegible, handwritten scrawl? This is a sign of an unfocused, lazy mind.

Try paraphrasing what you have written about each key point until you cut it down to a page or even a paragraph. Then repeat the process over and over until you can cut it down to a single short sentence. This is your key point. Some key points require this boiling down process; others just leap out and hit you.

Let's return to our example. Preparing for my presentation on vinyl Macintoshes, I used brainstorming to generate several key points. One key point revolved around the importance of permeability. I wrote down what I wanted to get across:

> *Our team here at the Department of Rainwear Investigation and Precipitation (DRIP) based its study on 99 volunteers divided into three groups of 33 each. Group A wore rubber boots and rubber raincoats, Group B wore vinyl boots and vinyl raincoats, and Group C wore gaberdine coats and running shoes. We placed all three groups in our patented storm simulator and left them to rot for one hour. After this procedure, we removed each individual's raingear and footwear and measured sweat concentrations using iontophoresis. This was a proxy measure of how much water had penetrated the rainwear and how much subjects sweated while wearing the raingear. We found that rubber-clad subjects had the highest sweat concentrations, gaberdine wearers had the least. Vinyl-clothed subjects were somewhere in between.*

Can we condense this paragraph into a single key point? Although the paragraph describes the study's methodology in detail, the most important part was the conclusion. What did the experiment show? "Rainwear must be water-impermeable but allow the wearer's skin to breathe; vinyl best fulfills that requirement." That sentence is still too long. Put the idea in point format, and you get, "Vinyl: most comfortable." That's the key point. Of course, later I'll build on this point by providing details and explanations about how I arrived at this conclusion.

Some people find it immensely hard to abstract in this way. To polish your skills, practice paraphrasing. I do it all the time. Many of the memos I receive are longwinded, unfocused, confused, and confusing. I often have to reread them several times before I have any idea what they're about, what they wish me to do, and whom they are from. If you can paraphrase a memo like that, paraphrasing your own key points will be easy.

When it's time to deliver your last key point, do so with extra zest, energy, and enthusiasm. Use high-energy action verbs. With a final startling statistic or statement, wake up audience members whose minds have drifted. Or actually state that this is your last key point. Knowing that the end is nigh, those who have not previously been riveted to your every word will rouse themselves from their stupor and focus their mental energy for this last hill. Knowing that it will all soon be over revitalizes audience members like no other energizing technique.

The Eyes Have It

Have you ever been watching the TV news when the Teleprompter suddenly stops working? You can tell right away, because the newscaster immediately looks down and fumbles with his or her notes. It's disconcerting, isn't it? Because the newscaster is no longer making eye contact with you, the connection has been broken and the flow has stopped. Even though the newscaster may be thousands of miles away from you, it's still important for him or her to look up at you while telling you the news.

What is true for the newscaster is true for you. Don't stare at your feet or your notes or the overhead projector while you make your presentation. Instead, practice the art of scanning. Look at a person for a few seconds, then move on. Be sure to include any people sitting behind you, even if you have to strain your neck to do so. Scanning includes everyone, but also demands attention from everyone. Too often presenters address their remarks only to a meeting's chairperson, even when the remarks are general in nature and applicable to all members. When you fail to look at every member of the group in turn, you exclude some and lose others' attention.

If you're a bit nervous or shy, you might find it difficult to make eye contact with members of the audience. Try looking at people's foreheads. No kidding. Work your way around the room spending five to 10 seconds looking at each forehead. Each person will think you are looking into the eyes of the person behind him or her. Continually scan the audience. Don't stare. Don't let your eyes dart around. And don't try this technique during a one-on-one conversation. The person you're talking to will be convinced there is something stuck to his or her forehead.

Golden Silence

You may only have 10 minutes for your presentation, but don't feel compelled to fill every second of it with your voice. Silence is the aural equivalent of white space. Ever noticed how a document with bullet points or paragraphs is much more attractive and easy to read than a document with densely packed, contiguous text? Silence does the same for the spoken word.

Used with care and moderation, silence can demand attention, differentiate your presentation's major parts, and mark milestones on the map of your presentation's logical flow. Silence is a wonderful way of recapturing lost minds, the people whose thoughts have drifted away from your presentation and who hear your voice as a distant drone. When the droning suddenly stops, the silence grabs their attention. Your silences do not have to be long. A few seconds can seem like an eternity. Don't overuse silence, but do add it to your toolbox of presentation techniques.

The End

Symphonies often end with a bang rather than a whimper. Movies end with a twist, a surprise, or a memorable line like "Tomorrow is another day."

Presentations are no different. They shouldn't just fade away into embarrassed silence. Unfortunately, they often do. I've attended many a presentation where the presenter suddenly says "...and that's it." The presentation just ends, without any sense of closure.

Good presentations have definite, clearly articulated endings. In your presentation's final phase, show a slide or overhead of a map of your presentation's logical flow. (See "Mental Maps" below.) Summarize your key points. Then use a summary slide or overhead to restate and reinforce your thesis sentence. That's the take-home message you will implant in your audience's minds. Remember the underlying three-part structure: Tell them what you are going to tell them. Tell them. Tell them what you told them.

Question Time

During your presentation, you are in control. Once question time begins, you are *still* in control. If you're lucky, you'll have a good facilitator who can orchestrate the question period for you. If your facilitator is ineffectual, don't be afraid to take over.

In an ideal world, presentations would be interactive tutorials where participants could ask questions whenever they needed clarification or reinforcement. That kind of interactive process works well at poster sessions, but the free flow of questions isn't always possible during oral presentations—especially during the 10-minute presentations that are popular at conferences and meetings. In most situations, you must relegate questions until the end to ensure that your presentation stays on track and on time. If you do allow questions during your presentation, be on your guard and keep an eye on the time. Have a clearly defined period of time for questions and do not exceed it. (If people still have questions, meet with them afterward.)

Always paraphrase a question before answering. By paraphrasing, you 1) clarify and simplify the question, 2) confirm that you understand the question correctly, 3) allow the rest of the audience to hear the question, and 4) gain vital seconds during which you can consider your response. Answering a question that nobody else heard is a surefire way to lose the audience's interest just when you need that interest most. Remember that unless microphones are provided for the audience, people sitting behind the questioner probably cannot hear the question, even if you can hear it perfectly.

Don't allow one person to monopolize the question period. Politely tell the person that you would be glad to meet with him or her afterward, but right now you would like to give others a chance to ask questions. Similarly, it's perfectly reasonable to interrupt someone who is actually making a long rambling statement instead of asking a question. Politely interrupt and ask the person to rephrase his or her statement as a short question, pointing out that time is limited and that others are waiting for a chance to ask questions.

Occasionally you'll want to turn a question back onto the person asking it or ask the audience's opinion. These techniques can be particularly useful if the issues are contentious or you need a few seconds to consider your own reply. A little humor can sometimes diffuse tension. But be very careful! Don't make anyone in the audience the butt of your jokes, even if they are being a pain in the neck.

Many people dread question periods and worry that they won't know the answers or will be badgered by an audience member. Don't panic! The vast majority of questions are straightforward and require straightforward answers. Do your homework. Keep up to date on the latest developments in your topic area. (Don't mindlessly recycle last year's presentation without checking that everything you said then is still true or applicable.) And try to anticipate questions by asking a colleague to listen to your presentation ahead of time and come up with likely questions. Taking these precautions will help ensure you can answer most questions.

If you don't know the answer to a question, say so. Don't be tempted to make up something. You are there to make a presentation, not to serve as the fount of all knowledge.

Besides, certain questions by their very nature may be unanswerable. They may require further research or be so complex that they're beyond the scope of the present discussion. If you find yourself drawn into a philosophical dialogue, offer to talk with that person one-on-one after the presentation.

Occasionally you may get hostile or even loaded questions. Avoid being sucked into internecine conflicts by refusing to comment on such matters. Don't meet hostile or sarcastic tones with like behavior. Watch your tone of voice and your body language. Always be polite, yet firm, in your reply. Never be rude or condescending. Never put anyone down, no matter how inane a question may appear to you. Treat everyone with respect. Be reasonable, but not a pushover. The key is to allow others the opportunity to save face. Grace under pressure is the hallmark of self-assurance.

Details

Effective presenters use statistical methods to summarize data, then use those summary data to show trends and correlations. Don't bombard your audience with the details. Relegate them to a supplemental handout or notes. Then make them available *after* the lecture. If you distribute handouts during your presentation, everyone will stop paying attention to you to read them. Of course, if you want people to annotate their handouts, you'll need to distribute them before your presentation begins.

Stay Late

Why not just run out as soon as your standing ovation is over? Staying around after your presentation shows respect: respect for any speakers who follow you, for the organizers who invited you, and for those who came to listen to you. No matter how much you wish the day were over, never hit and run.

Staying late also gives your audience an opportunity to explore points of particular interest arising from your presentation. Such discussions don't just benefit your audience: You may learn about the latest findings or newest initiatives. Remember that we have twice as many ears as mouths. Use them! Talk to people, but be sure to listen, too.

Staying behind also allows you to get a feel for how well your presentation went. Sometimes it's not possible to use formal feedback procedures, such as handing out anonymous evaluation questionnaires. Chatting with people after your presentation is a subjective and limited way of obtaining feedback, but it's better than nothing.

The Uses of Language

Think carefully about your mode of speech. Here are some points to consider:

- Use action verbs whenever possible. They transmit energy, vitality, and enthusiasm to your audience.
- Watch out for common mistakes. The commonly used word "irregardless," for instance, is not an actual word. Other words get mixed-up a lot. Contrary to popular opinion, "disparaging" and "disparate" have never been synonyms. Gaffes like these undermine your credibility and ruin an otherwise excellent talk.
- Avoid long or obscure words as much as you can. Because science seems to thrive on incomprehensible technobabble, it's not always possible to follow this rule. If in doubt, use the shorter, simpler word.
- If you are going to use a technical or unusual word, be absolutely sure of its meaning.
- Beware of malapropisms, such as saying "Avery" instead of "ovary" or "gambit" instead of "gamut."

Jargon

Verbosity is not a sign of intellect. It is a sign of poor communication skills. Be sure that any jargon—as well as abbreviations and acronyms—are commonly understood or clearly defined. This is especially important for poster presentations, because 95 percent of the people who visit your presentation will not be experts in your field. Your jargon, acronyms, and abbreviations may mean something very different to them. One way to be sure that you're speaking the same language is to have someone who is not an expert in your field and who is not familiar with your work review your presentation while you're working on it.

Analogy, Metaphor, Allusion

When you're planning your presentation, turn to analogies, metaphors, and allusions whenever you need to introduce unfamiliar concepts. These figures of speech can be powerful tools in your pursuit of clarity and brevity. But remember, not everyone in the audience will be familiar with concepts that seem obvious to you. To avoid confusion, keep things simple.

Use analogy to explain esoteric matters. The technique works by likening an unfamiliar concept to a familiar one. You could explain the immune system to a lay person by comparing it to a police force, for instance. The immune system patrols the body; identifies bad guys, such as cancer cells, foreign cells, or viruses; and mobilizes help to neutralize or destroy them.

Be sure to use an analogy your audience can understand. There's no point in using one misunderstood concept to explain another! For example, don't try to explain the nervous system's functioning to a lay audience by comparing it to the "bus" on a CPU's motherboard, which receives and transmits data bits between various peripheral devices and the microprocessor. Computer science students may understand this analogy, but most other audiences would be confused. A better analogy would be the telephone system, which transmits messages from a central office to various branches. The telephone system is almost universally familiar; the architecture of personal computers is not. Choose analogies related to something in your audience's everyday life or work. And beware of overworking your analogy.

Use metaphors when you want to express an idea succinctly. Sports metaphors are particularly commonplace. We speak of "ballpark" figures, being a "team player," or doing an "end-run" on our boss. Be careful not to mix your metaphors: "Before we bought the new bar coding module for the Pangalactic Hyperwhiz Omnianalyzer, we were drowning under a mountain of paperwork." You can drown in a sea of paperwork or be buried under a mountain of paperwork, but you can't drown under a mountain. Used with discretion, however, metaphors will serve you well in your quest for effective and efficient communication. Like pictures, metaphors can save a thousand words. So can allusions, such as describing something as being "Mickey Mouse."

Don't overuse such devices. It is all too easy for them to become tired, trite clichés. And although analogies, metaphors, and allusions can help us bridge gaps in understanding, they can also act as barriers to communication. If your audience doesn't understand the allusion or metaphor you choose, you have compounded the communication problem by introducing yet another unknown concept.

Be especially aware of these potential problems if you give presentations to international groups, whether you're in your country or theirs. Analogies, metaphors, and allusions are only meaningful in the context of a shared culture. For example, Britons do not use the phrase "Mickey Mouse" to describe something very easy the way we do. In France, Mickey is not even called Mickey. Even within your own country, certain analogies, metaphors, and allusions may be in bad taste. Be particularly wary of figures of speech that arise from or perpetuate cultural or racial stereotypes. We often use these terms unconsciously, not meaning to offend.

Clichés

Avoid clichés. Although they can be useful, their triteness tends to bore peoples. If you'll pardon the cliché, familiarity breeds contempt.

Moreover, clichés—like analogies, metaphors, and allusions—assume a common knowledge base in the audience. The audience may understand English but not all of its idioms. Even if there is a Russian, Japanese, or Icelandic equivalent of "Too many cooks spoil the broth," the audience may not understand your allusion. Similarly, an Islamic or Jewish audience may be unfamiliar with the sermon on the mount, the beatitudes, or the fact that Saul and Paul were not two different people.

Friends! Romans! Countrymen!

Winston Churchill spent a great deal of time crafting his speeches. He would play the music of language in his head, like a poet in search of the perfect rhyme. He would try out phrases on friends and colleagues. He would use the rhythm of a three-part phrase: "Never in the history of human conflict was so much owed by so many to so few."

Lincoln was another powerful speaker. Everyone has heard of his Gettysburg address, but no one remembers the two-hour speech that preceded it, which was given by a "professional orator." Lincoln's speech was so short that the photographer almost missed getting a picture of the President making it. It was widely ridiculed at the time as "unpresidential," "unfitting for so great an occasion," and "too brief by far." History has been the judge, however.

In a few short sentences, Lincoln said it all. The speech remains a masterpiece of brevity and clarity, full of poetry, truth, and the music of the English language. It flows with phrases that ring out through the ages: "all men are created equal," "the last full measure of devotion," and "government of the people, by the people, for the people." Lincoln's speech did have one error: Its author claimed, "The world will little note, nor long remember what we say here."

I am not saying that we must all be like Mark Anthony,[7] Lincoln, or Churchill, great orators who stirred their audiences' hearts and minds to heights of outrage or passion. I merely draw your attention to the inherent beauty of the English language and its power to persuade. Therein lies the essence of what makes one person's words memorable, while another's are lost forever: Great speakers use rhetorical devices.

Equipment

Even the best presentation in the world can be spoiled by the inappropriate use of equipment.

Anchors Aweigh!

Ditch the lectern whenever possible. Speakers often hide behind their lecterns. Instead of a place to put their notes, the lectern becomes a shield. Short speakers will find themselves peering over the lectern's top. Nervous speakers will find themselves drawn to the lectern as a safe haven that protects them from the audience. Nervous speakers trying to pretend they are relaxed will lean on the lectern, put one hand in a pocket, and give the entire presentation in a nonchalant but incredibly dull monotone without ever moving from the spot.

Lecterns are psychological barriers between you and your audience. They also act as anchors, inhibiting your movement. It's difficult to convey excitement or enthusiasm if you spend the entire presentation glued to the spot behind the lectern. To keep the audience's interest, move around the stage. Don't pace up and down like a caged animal, but use your entire body to communicate your message.

[7] A good example of an allusion! The allusion is to Mark Anthony's famous speech in Shakespeare's *Julius Caesar*. Mark Anthony incites the people of Rome against Caesar's murderers with a masterfully ironic speech.

Freeing yourself from the lectern may require a little advance planning. That's because static microphones force you to stay rooted behind the lectern. Request a clip-on or portable microphone well ahead of your presentation day. At the very least, remove the microphone from its stand and carry it around with you if possible. Some speakers actually carry their own radio microphone and receiver along with a set of miscellaneous connectors and adaptors to ensure that their microphone can be plugged into any auditorium's amplifier system.

You may be wondering where you're supposed to put your notes if you can't use the lectern. What notes? Truly effective presenters don't use notes. Truly effective speakers know their material forwards, backwards, inside out, and sideways in Babylonic cuneiform. Presenters who don't know their subjects that well should not be presenting.

Effective presenters have prepared and practiced carefully. They have mapped out their presentations. They use overheads or slides as their cues. Nothing is as unconvincing, monotonous, and dull as a presenter reading from notes. Just ask any politician's spouse. Constantly looking down at notes also distracts you and keeps you from maintaining eye contact with your audience. The result? Your presentation will lack credibility. Your audience will be thinking, "If she's the expert, how come she has to keep looking at her notes?"

Jedi Knight Syndrome

And what of the laser-light pointer? There are many would-be Luke Skywalkers out there, who love to whirl their pointers around. They accidentally point them into people's eyes. They induce nausea, vertigo, and projectile vomiting in all but the most comatose of audiences. And they distract everyone from the messages they are trying to get across.

Remember Yoda's advice to the Jedi knights: Use The Force. Better yet, use an old-fashioned pointer. You can buy telescopic pointers that retract into themselves until they're as small as a pen. The advantage of these is that you aren't tempted to whirl them round and round as you highlight your data. Plus, you must move to the screen to use them. That movement provides interest and helps emphasize your points better than any whirling laser-light pointer can. Because your entire body is involved in making the point, your presentation becomes an active process. (Keep in mind, however, that your hand movements near the projector are magnified disconcertingly. Lay a pen or pencil on the projector screen and use that as a pointer.)

Sometimes you can't avoid laser-light pointers. That's especially true in very large lecture halls where the screen is mounted high up. On these occasions, use the laser-light pointer only when you need to. Don't whirl it! If you do, neither The Force nor the audience will be with you for long.

Let There Be Light!

Have you ever been at a presentation where the speaker switches on the projector and switches off every other light in the auditorium? This allows a speaker to hide, to be a voice in the shadows. It also puts most of the audience to sleep.

Set light levels no lower than is absolutely necessary to ensure that projected images are clear. (Familiarizing yourself with light switches and dimmer controls is yet another reason to arrive at your presentation venue early.) If you can't control the light level yourself, recruit someone to help you or learn how to signal the auditorium staff. You have invested a considerable amount of time and energy preparing for your presentation. Don't allow inappropriate light levels to ruin it.

A judicious use of light can enhance your presentation by punctuating your talk. Turning off the projector and turning up the lights can help focus the audience's attention back on you, signal the end of one section, or move the audience with you onto the next concept. A compelling talk is like a symphony, with changes of tone, mood, and tempo. We like contrast. Our brains eventually tune out constant droning noises, such as monotonous speakers. Changes in lighting, just like changes in your voice or position, help you maintain audience interest and attention because they provide contrast. Don't overuse this technique, however. Changes in light levels should be occasional punctuation marks, not strobe light shows that induce epileptic fits in your audience.

Summary

This section on oral presentations began by discussing methods for overcoming stage fright and building credibility. The section then dissected the anatomy of an oral presentation, from establishing a word budget to ending with a flourish. As we learned, an oral presentation begins with topic and thesis sentences. The middle section consists of key points in reverse order of importance. The end features a summary and take-home message. Part Two also included sections on how to use language and equipment in ways that enhance rather than detract from your presentation. In Part Three, we'll learn how to make the most of visual aids.

■

PART THREE: VISUAL AIDS

Part Three: Visual Aids

Bridging Parts Two and Four, this section discusses visual aids suitable for both oral presentations and poster sessions. The section begins by describing the surprisingly powerful effect visual aids can have on things like audiences' attention levels and their perceptions of value. The section then discusses various ways you can use visual aids, such as mental maps or slides of key points, to enhance your presentations. A series of graphs illustrating right and wrong ways to produce visual aids is the highlight of a section on design principles. The section concludes with a brief look at electronic presentations and handouts.

The Effects of Visual Aids

Visual aids do a lot more than simply give your audience something to look at while you talk. If they're well-designed, they can have the effects described below.

Attention!

Effective visual aids demand attention. They are like beautiful people: You cannot keep your eyes off them. Attract viewers' gaze, and you grab their attention. Grab their attention, and you capture their minds. Capture their minds, and you can implant your message.

Why are visual aids like slides and overheads so effective in enhancing our ability to communicate? We are visual beings. Visual aids increase the average audience member's attention span by eight percent. They increase comprehension by nine percent. They increase retention by 10 percent. The simple act of putting an idea onto a slide or overhead automatically increases the likelihood of that idea being noticed, understood, and retained.

Using high-quality visual aids helps ensure that you will leave a positive, lasting impression with your audience. Remember, success grows exponentially. Good presenters are asked back and are asked to give new presentations elsewhere. Good presentations open doors to all kinds of opportunities.

Perceived Value

Manufacturers often make cheap briefcases of simulated leather. Taking advantage of leather's association with luxury and expense, they thus increase their product's perceived value.

Two restaurants may have more or less the same quality food, yet you prefer one to the other. Why? Because its extremely polite waiters and waitresses enhance its perceived value.

A well-dressed, goodlooking used-car salesman with a smile and silver tongue is far more likely to persuade you to buy a car than is a scruffy, mumbling, dour person. The actual value of these two people is unknown. The scruffy guy may be a wonderful human being, while the goodlooking smoothtalker is an evil megalomaniac intent on world domination.

Unfortunately, it is perceived value rather than actual value that counts in our cursory interactions with others. That may seem shallow. It is. But it's the perceived value that counts. The same phenomenon is at work in presentations. You can boost the perceived value your audience finds in your presentation by 11 percent simply by using well-designed overheads or slides. A shallow but nonetheless true fact of life.

Persuasion

Visual aids increase your powers of persuasion by six percent. That could be crucial in situations where you are trying to persuade someone to choose one product or course of action over another. Say your boss is planning to introduce some changes in the department. You have an alternative plan, and she gives you a chance to present your proposal. You use carefully designed overheads or slides to give yourself a competitive edge: That extra six percent of persuasive power ensures that your ideas prevail.

You should exploit anything that boosts your persuasive power. Even if you're just presenting pure research, you are still trying to persuade. You are trying to persuade your audience that your work is valid, trustworthy, and significant—and that you are, too. Persuasion is the art of gaining entry into a previously closed mind and infecting it with your ideas.

Action

Using visual aids increases the likelihood of spurring people into action by a staggering 43 percent! Think about this. Suppose you are trying to persuade your boss to change the current shift schedule. You are 43 percent more likely to spur her into action if you show her a well-designed chart outlining your proposals than if you simply stroll into her office and say, "I think we should change the shift rotations."

Whenever you want someone to do something—whether it's a board that needs to take action, a committee that needs to change its recommendation, or an administration that needs to change its policy— incorporate well-designed visual aids into a presentation. You'll not only persuade. You'll see some action.

Professionalism

The term "professional" is much overused and abused nowadays. At one time, the term was clearly understood to include fields like medicine and law. Now we have professional sales associates when we used to have sales assistants. We have professional sanitation consultants when we once had to make do with sewer workers. We have professional estheticians, although in the past we somehow managed to get by with hairdressers. Today everyone is a professional.

That doesn't mean everyone acts professionally. When I use the term "professional," I mean things like high-quality, competent, self-confident, and well-respected. Being regarded as professional in this sense is a total package. It involves people's perceptions of your worth, your performance, your qualifications, and your experience. By using carefully prepared visual aids, you will build and enhance your professional image.

Interest Pays

We are very visual beings. We like to keep our optical nerves busy. That's why carefully prepared visual aids help you seem more interesting on a subtle, psychological level. The simple act of looking at a concept on a slide or overhead increases your audience's interest.

Why is it important to generate and maintain interest? Surely if people come to your presentation voluntarily they must be interested, right? Not necessarily. Students may be required to attend your lecture whether they're interested in your topic or not, for example. Some people may attend your presentation because everyone expects them to or because they want to impress their boss. Even at scientific meetings where the audience genuinely does want to be there, it is still necessary to generate and sustain interest. People are not machines. They tire. They get irritable. They have limited attention spans and cannot absorb information forever. If your presentation is the fifth one of the morning, they have probably lost some of their ability to concentrate. They may be hungry, sleepy, or hot. Give them something to look at, and you'll keep them from slipping into stupors.

The Uses of Visual Aids

Mental Maps

Excellent researchers often make very poor presenters. Because their research is highly specialized and narrowly focused, they sometimes find it difficult to step back from their small corner of the cosmos. They launch into the arcane details of their topic without first setting the scene. They forget to explain how their research relates to other areas. They tell us in exquisite detail about a needle they found in a haystack, but forget to tell us why the needle is important or why they were looking in a haystack in the first place. Instead of helping their audience across the stepping stones to the other side of the river of understanding, these presenters simply assume that everyone is across the river and ready to journey into the land of Esoterica.

When you give a presentation, remember that your audience doesn't know where your talk is leading. They are traveling through a country where they have never been before. They want to know where they're going and how they're going to get there. Without a map, many audience members will be disconcerted.

Ensure effective communication by showing the audience a map outlining your presentation's logical flow—where you are going and how you're going to get there. Think of each section of your presentation as a milestone along the journey. At each of those milestones, show a slide or overhead with a map to show where you have just come from and where you are going.

In fact, use maps showing your presentation's logical flow whenever you can. Attention drifts. Use maps to help get your audience oriented again. If someone has gotten lost, your map will help them rejoin you on the rest of your logical journey.

Mental maps can help you, too. As you create them, you are forced to consider carefully your presentation's logical structure. You are forced to think about the progression of ideas that you will present. In short, you are forced to plan your journey carefully. As we all know, unplanned journeys often end in disaster. Prevent disaster—and maximize clarity and accessibility—by mapping your presentation. Then enjoy the trip!

Key Points

Humans have a limited capacity to remember new things. We are not the organic equivalent of audio- or videotape recorders. We cannot simply dump information into our brains. Fortunately, there is usually a handful of facts that represent a presentation's essence. Highlight those key points both orally and visually.

Let's say you are presenting your fabulous research on the impact of rain showers on people. You could show dozens of graphs and tables as you describe your methods and the dozens of different kinds of umbrellas, raincoats, and boots you used in your experiments. Or you could summarize this mountain of information by showing a slide of a drenched man. The take-home message? If you don't wear raingear in a rain storm, you get wet.

Visuals help your audience understand and remember that take-home message. In fact, a well-designed graph can save a thousand words by showing a correlation or trend at a glance. In turn, a single sentence can summarize the graph. The process of simplification distills your message's essence from the raw data to the graph to the single statement that represents your key point.

Concepts and Diagrams

Whenever you are trying to get a concept across to an audience, remember the old saying: A picture is worth a thousand words. So is a diagram or a drawing. When you're planning your presentation, consider whether key concepts could be better explained using a diagram or drawing rather than words alone.

Want to understand how audiences feel when presenters explain unfamiliar ideas without using visual aids? Try this test. Draw 18 lines on a piece of paper as follows:

- Make 12 lines that are each seven-eighths of an inch long.
- Join them at right angles to each other to make three individual squares.
- Join the remaining six lines into sets of two each, joining them end to end to form three individual right angles.
- Join the two outside points of each right angle to the upper corners of each square to form three separate pentagons.

Figured it out yet? If so, how many neurons did you wipe out in the process? If not, turn the page for a picture of the results. Now you know how audiences often feel.

As you may have noticed, it's almost impossible to understand complicated ideas if words are all you have to go on. Without visuals, your audience has to do a lot of mental work to develop their own images of what you're describing. Don't make your audience's brains hurt. Use a diagram. Your audience will understand your idea immediately.

Visual metaphors or allusions make things even easier for your audience. Imagine that you're trying to explain genetics to a lay audience, for example. They have no idea what a gene is, and they do not speak Genetish or Molecularbiologish. Reinforce your explanations with metaphorical overheads or slides. When you're explaining what a gene is, you might show a set of blueprints. Be sure you limit yourself to one idea per slide. Any more than that will reduce the analogy's impact.

Don't try to cram too much onto each slide or overhead. Visual aids that contain too much information overwhelm rather than inform. They are impediments to learning, rather than aids. To audience members, data-intensive slides become no more than a blur of numbers. You'll lose their attention. Worse, you'll cause them to lose their collective consciousness. Use visual aids to extract key data. And keep in mind that most audiences cannot read text smaller than 18 points.

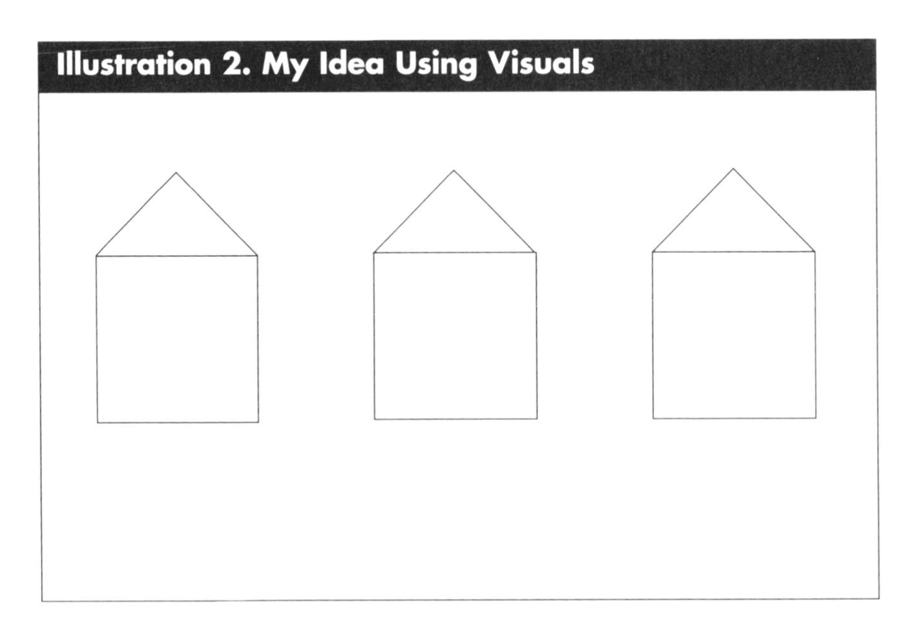

Of course, you don't need visual aids to transmit every idea. Occasionally words are all that is necessary. Indeed, visual aids can sometimes detract from the natural flow and render the sublime ridiculous. To be effective, visual aids must be used with care and discretion. Some presenters have so many slides that they might as well have made a movie. As a general rule, use one visual aid for every two minutes of presentation time.

Design Principles

In his book *The Visual Display of Quantitative Information*, Edward R. Tufte identifies the principles of good graphic design. The book is a superb introduction to the pitfalls of graphical representation. Of course, he also tells you how to do things right. His approach boils down to four simple rules:

1. Above all else, show the data.
2. Maximize the data-to-ink ratio.
3. Erase nondata ink.
4. Erase redundant data ink.

Which Format?

The next time you need to present your data, review your graphs carefully. Always ask yourself three questions: 1) Is a graph the best way to present my data? 2) What type of graph is most appropriate? and 3) Have I applied Tufte's rules? Then consult the table below.

Type of Graph	Use	Suggestions
Title	To introduce your presentation or new topics within your presentation	Keep titles short, using no more than eight words.
Table	To show statistical, aggregate, or simple comparative data	If possible, limit the number of rows and columns to seven or fewer. Group smaller categories into a category called "Other."
Pie	To show relative proportions of a whole	Don't slice the pie into more than six pieces. Group smaller categories into a category called "Other." To emphasize an important slice, cut it away from the rest of the pie.
Vertical Column	To make comparisons that are time-dependent and categorical	Limit the number of columns to six or fewer. Group smaller categories into a category called "Other."
Horizontal Column	To make comparisons that aren't time dependent or categorical	Limit the number of columns to six or fewer. Group smaller categories into a category called "Other."
Line	To show trends over time	Limit the number of series to five or fewer. Use a thicker line or different color to show the most important line. Use broken lines to show extrapolations.
Area	To show changes in volume, totals, or quantity	Limit the number of areas to six or fewer. Put the lowest and flattest, or least variable, series at the front of charts that use an overlap style. Use stacked area charts to show cumulative totals.
High/Low	To show changes in range over a fixed period of time	Limit the number of series to six or fewer.
Scatter	To show correlation (or not!) between variables	Use line-fitting where appropriate, such as in regression analyses.

Graphs 101

The graph is perhaps the single most widely used, abused, and misused tool in scientific communication. Of course, graphs can be wonderful. As concrete pictures of abstract relationships between variables, they can save you a thousand words. But many scientists don't think twice about a graph's design once they've selected the right scale and plotted their points. Unfortunately, a poorly designed graph can mislead, misinform, or even lie.

Whenever you're designing or editing a scientific graph, watch out for common no-no's. Beware of what I call "chart junk," for instance. This junk comprises all the non-essential gridlines, boxes, and other items that often adorn graphs. Chart junk only distracts.

A related problem is what I call "artritis." The disease is characterized by graphs in which data are lost in a montage of clip art, three-dimensional effects, and other supposedly artistic efforts. Another symptom is a presenter's use of graphs that aren't even necessary in the first place.

Avoid moiré patterns. Often used as shading, these patterns lead to unintentional optical art. The pattern's parallel lines interact with naturally occurring tremors in the eye to produce an illusion of vibration.

Whenever possible, avoid using legends. Legends make viewers work harder to understand your graph: They have to study the legend, then examine the graph, and then check the legend again to be sure they haven't confused the short-dashed broken line with the long-dashed broken line or the asterisks with the double asterisks. In the meantime, they stop listening to what you are saying. A graph should augment the clarity of your talk, not diminish it. Don't make your audience take on mental work.

To see these suggestions at work, let's look at some graphs. Start with Version One, a magnificent piece of graphsmanship, packed with information, interesting typefaces, and cute clip art. It is a visual tour de force, a feast for the eye, a thing of beauty. Or is it? The graph may look impressive, but what does it actually tell you about the data? Without looking at the graphs on the following pages, think about how you might improve this graph.

Illustration 3. Version One

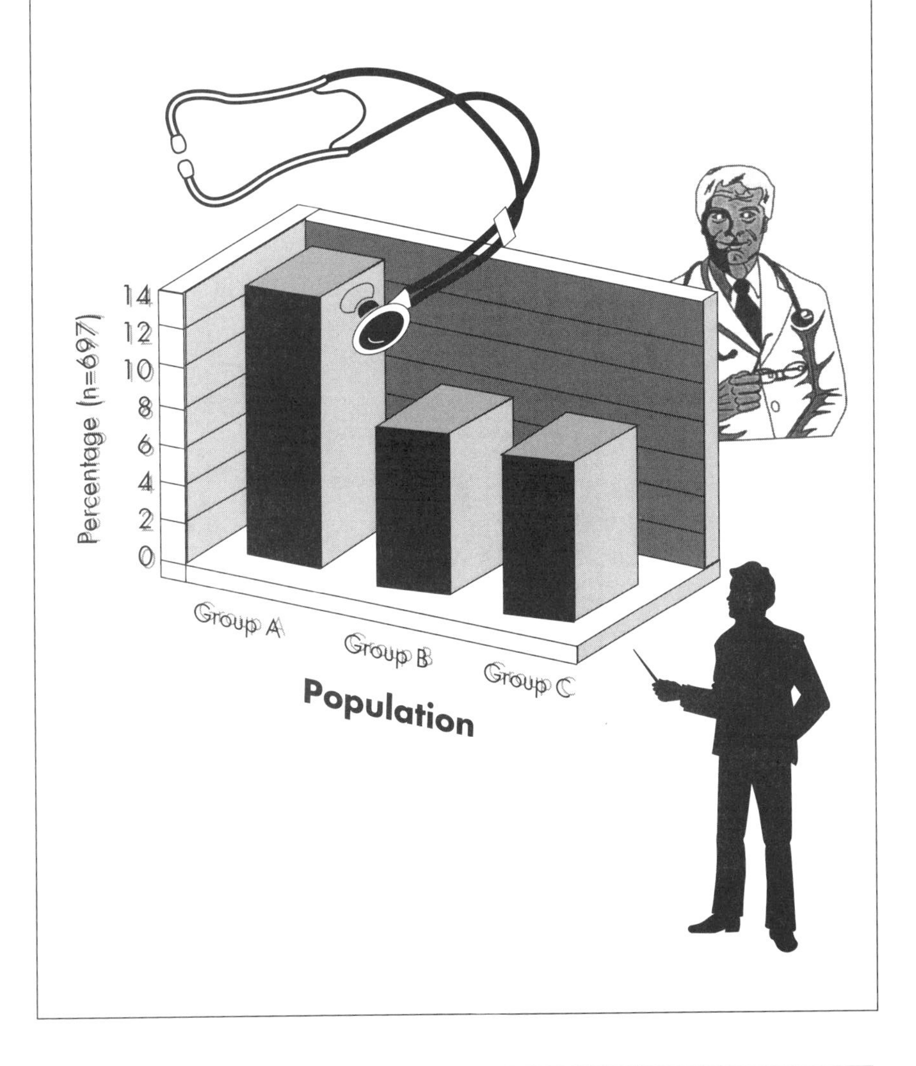

Now look at Version Two. What has happened? Well, the fancy box around the title is gone. So is the noisy background, which I have replaced with some nice straight lines corresponding to the ticks on the vertical axis. I've replaced the Gothic typeface with a much clearer one. And I've gotten rid of the cute clip art. It looked good but contributed nothing to the graph's clarity. In short, I've removed all sorts of things from the first graph. Things are looking good. Or are they? Look carefully at this "improved" graph. Again, list possible improvements before you peek at the next graph.

Illustration 4. Version Two

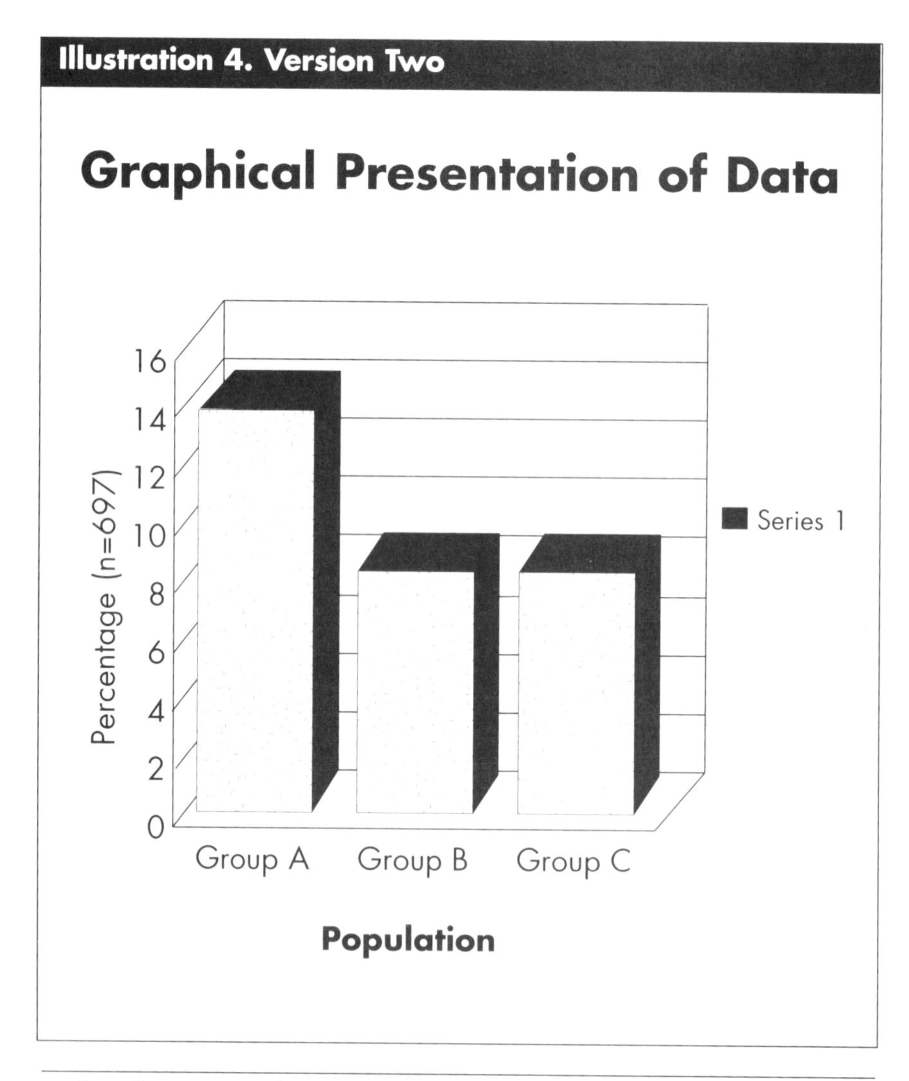

What's different about Version Three? The major change is that I've switched from three dimensions to two. Since I'm only presenting two variables, I didn't need three dimensions to present my data clearly. Computers have made it easy to produce fancy effects like three dimensions and shadows. Unless these effects increase your data's clarity, resist the temptation to use them. This version is looking good, right? Maybe not. Again, list potential improvements before looking at the next graph.

Illustration 5. Version Three

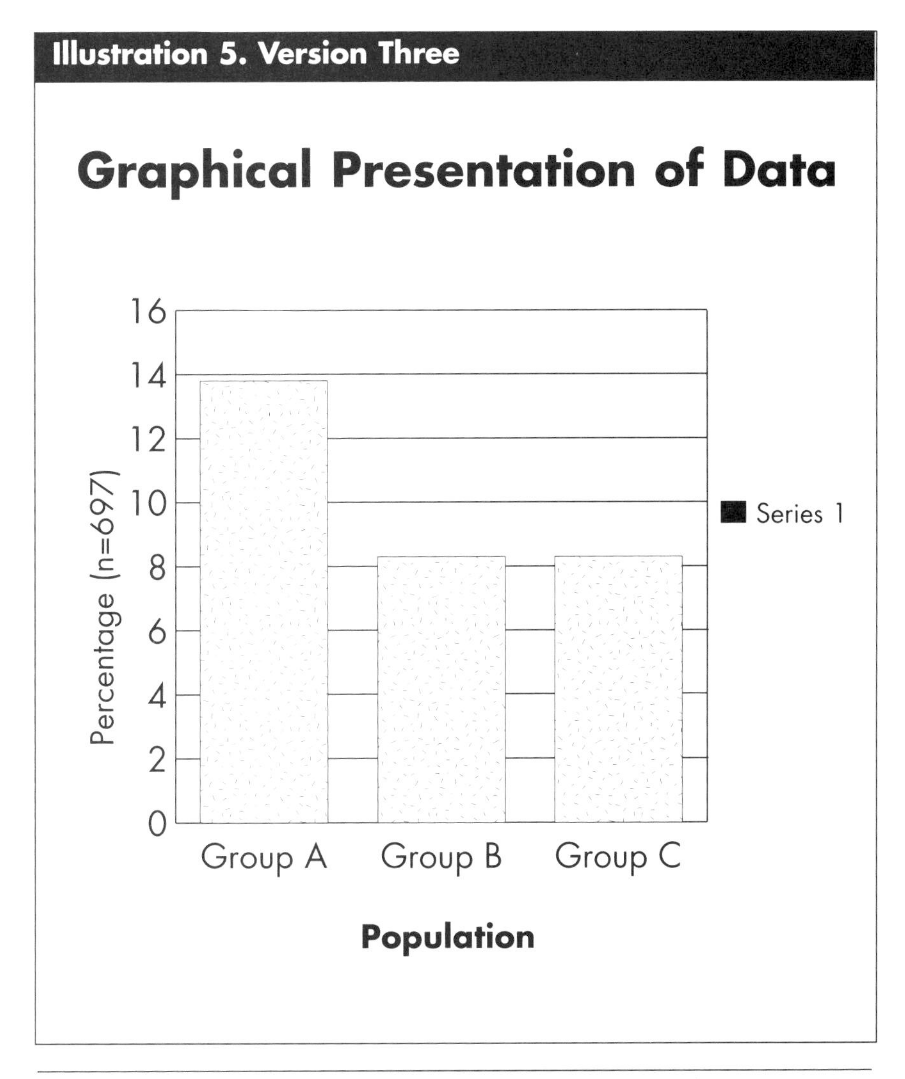

In Version Four, I've tackled the graph's vertical and horizontal lines in an attempt to make the graph less confusing. I've removed the background grid, for instance. Although horizontal lines can help viewers scan values along the "y" axis, in this case the lines were more likely to confuse than clarify. (If you use too many closely spaced parallel lines, you'll create optical effects that make the eye see curves that aren't really there.) I've also removed the graph's frame, which contributed nothing to the graph's message. I've even removed the lines marking the "x" and "y" axes, which were simply distracting. Any ideas for further improving the graph?

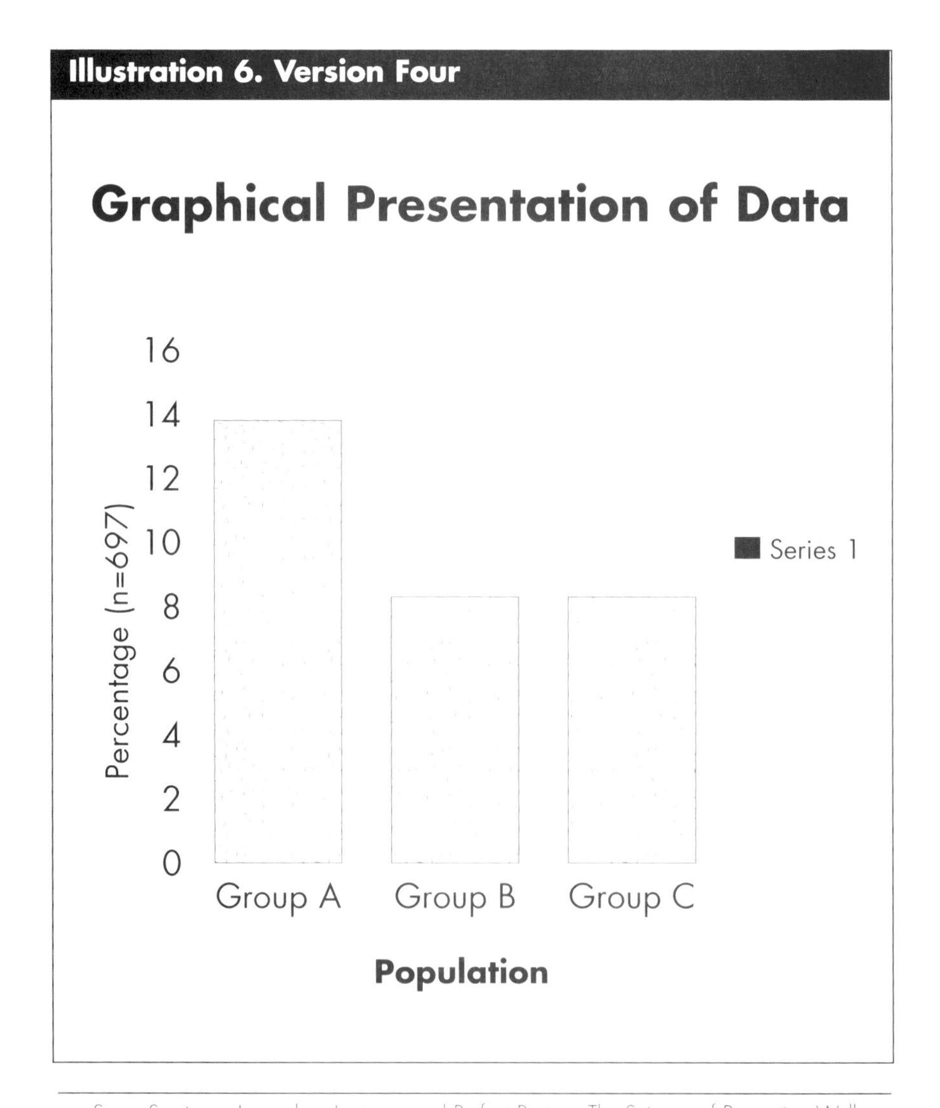

Look at Version Five. The major change is the shading of the columns. So what? Turn back to Version Four and stare at the shading for a few seconds. These so-called moiré patterns interact with your eyes' natural tremors to produce the illusion of vibration. That illusion prevents the eye from focusing properly, which is not only irritating but tiring. Unfortunately, moiré patterns are endemic in the scientific literature. Avoid them. This graph is a huge improvement over the original. Can you find any more ways of improving it?

Illustration 7. Version Five

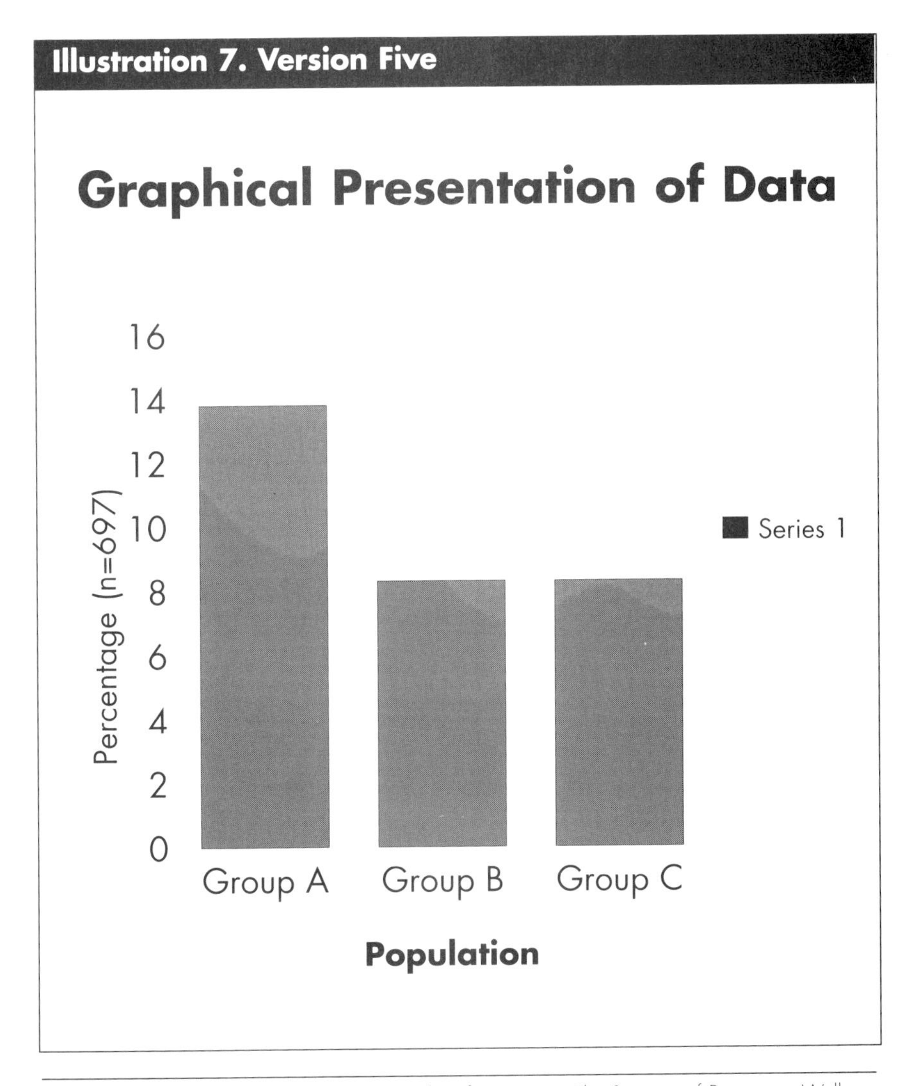

In Version Six, I've simplified the graph even more. I have removed the box announcing "Series 1," since there is only one series of data. And I have replaced gridlines with white space. The use of white space is a fundamental concept of graphic design. In the quest for clarity, we are too quick to add ink when we should actually do the opposite. Removing ink—creating white space—is a highly effective technique. In this example, I've simply inserted white spaces into the columns. This creates the illusion of gridlines without the confusion of black parallel lines. Perfection! Or is it? Look again.

Illustration 8. Version Six

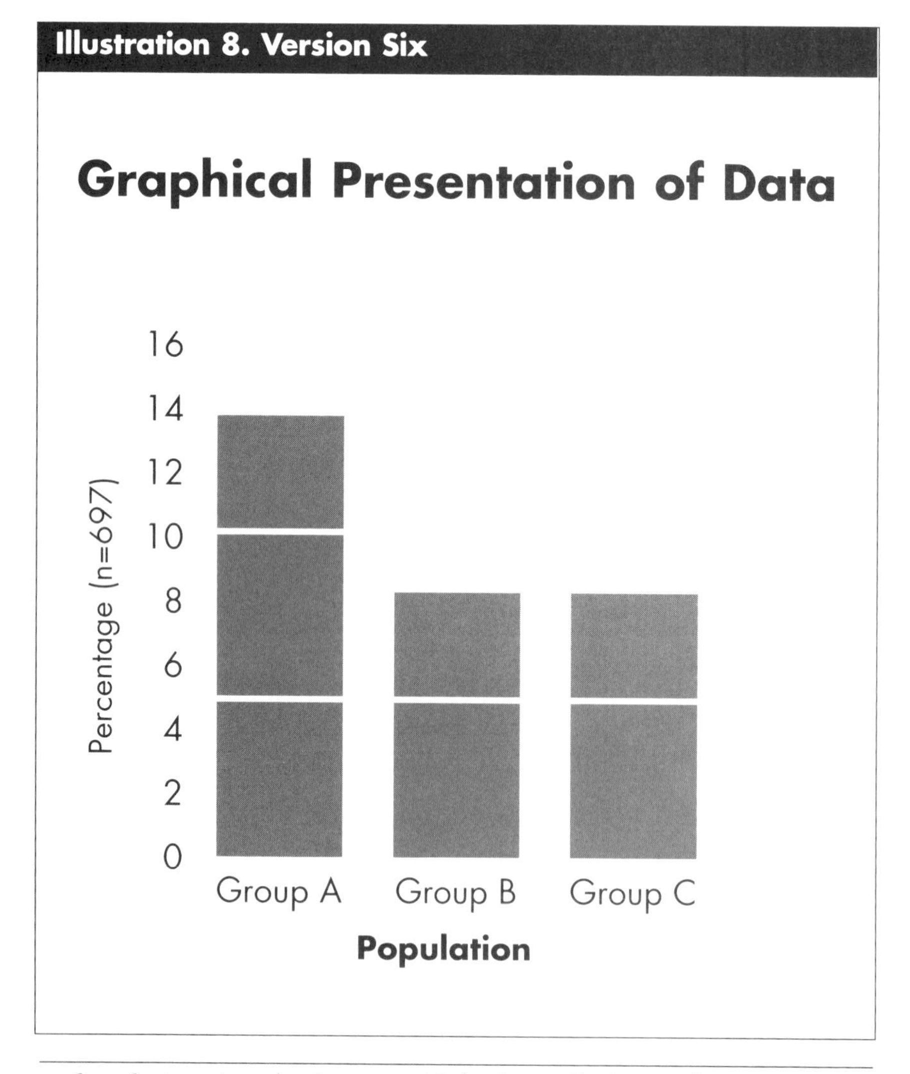

The change in Version Seven is obvious: I have used a completely different approach to presenting my data. By using a pie chart, I can show the relative size of each data category. Unfortunately, I have also lapsed back into an unnecessary three-dimensional graph. Some other graphical bloopers have also slipped back in. Can you spot them?

Illustration 9. Version Seven

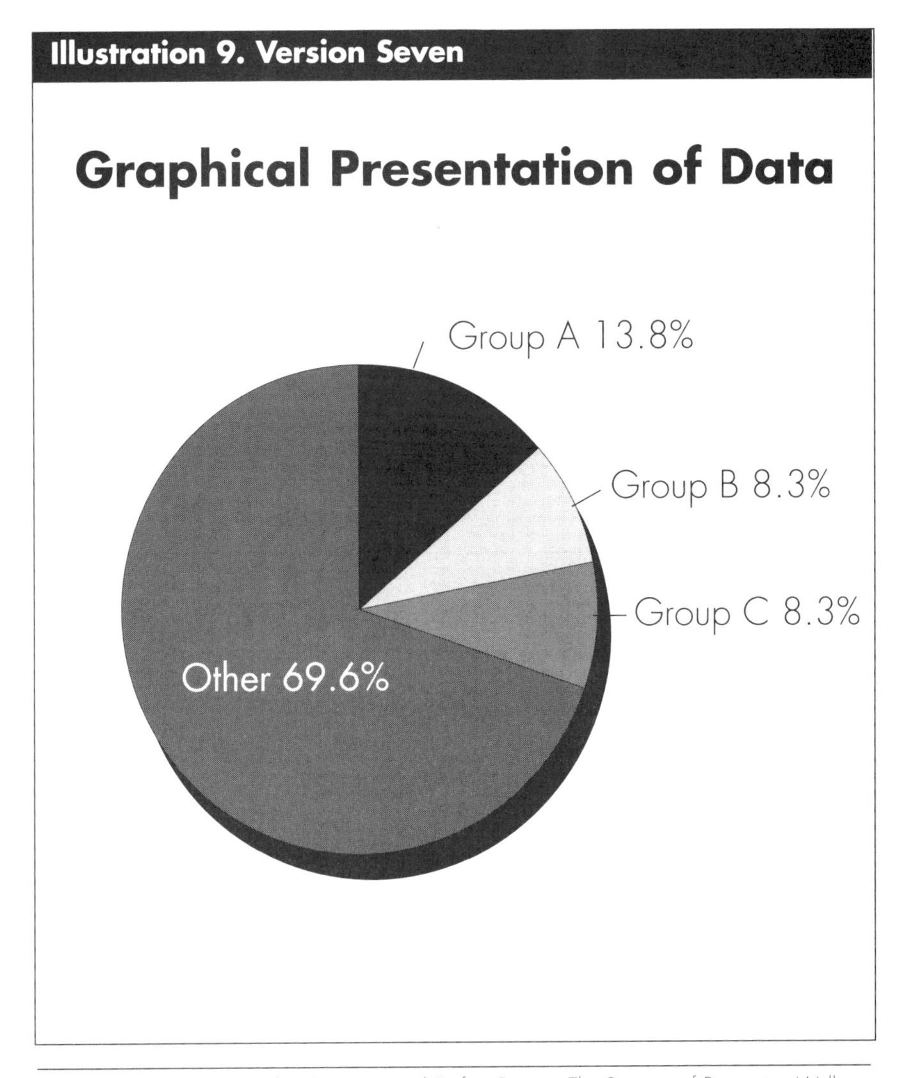

With just a few changes, we could end up with a perfect pie graph. But is a pie graph the best solution? Was a graph the best solution to the problem of conveying my data in the clearest way possible? That should have been our first question. If it had been, we could have saved ourselves a lot of work. Always make sure you are in the right forest before you start chopping down trees!

As you can see in Version Eight, a table was a better choice all along.

Illustration 10. Version Eight

Graphical Presentation of Data

Population	Percent (n=697)
Group A	13.8
Group B	8.3
Group C	8.3

Tables 101

As the example in the last section shows, tables can be an effective way to present information. Just follow these simple rules:

- Keep the number of rows and columns to an absolute minimum. Whenever possible, use summary data. Don't try to show everything. Remember, less clutter means more clarity.
- Highlight key information with a colored row or bold font.
- Keep headings brief and specific.
- Limit recurring units or symbols, such as "$" or "%," to the headings. Don't keep repeating them in every entry in every row.
- Group less important information in a general column called "Other." Better still, omit it altogether.

Color

A splash of color against a gray sky, a rainbow commands our attention after the rain.

Remember, we are visual beings. Much of our understanding of the world around us is based on the visual input of information. We see our world in a rich range of colors, hues, and tones. When we're watching a black-and-white movie or attending a presentation that only uses black-and-white overheads, we feel deprived.

Learn from advertisers, the masters of the message. They know that consumers are 80 percent more likely to read color ads than black-and-white ones. They know that color ads boost retention 78 percent over black and white. And they know that color ads increase sales by 85 percent.

These are astounding statistics: By using color, you can increase your viewership, increase retention, and increase the likelihood of making a sale. This is particularly important to know when you're trying to persuade people to act. You might be trying to convince your administration to pay for a brand-new Pangalactic Hyperwhiz Omnianalyzer or to reverse its decision on how to implement cost savings. As the advertising statistics suggest, color is a very effective means of increasing your powers of persuasion.

Color gives your presentations that special edge that differentiates you and your work from others. Used carefully, color adds a new dimension to the efficient, effective transmission of information. It's as if you suddenly had many more channels available to carry your message.

Like all powerful things, of course, color should be used with discretion: Garish slides or psychedelic overheads will have the opposite effect than intended. In addition to hurting the eye, they insult the audience's aesthetic sensibilities. And remember that color can also carry cultural meanings. In Western societies, for instance, black is the color of mourning and white the traditional color of weddings. In China, however, white is the color of death and red the color of celebration. Be aware of these differences when addressing ethnic groups different from your own or when giving presentations in other countries. The color of your slides, overheads, and even clothes may have meanings for your audience beyond anything you intended.

Color printers for personal computers are now widely available and reasonably priced. You can use a color inkjet printer to produce high-quality overheads directly. Or you can use your computer to design 35mm color slides, which you can then send via modem to slide-printing services offering rapid turnaround times.

Many computer graphics programs now include so-called "wizards," automated assistants that—theoretically—do much of the initial work of designing slides or overheads for you. Although wizards can be useful, be careful when using software packages designed for use in the fields of business or art. The graphs they generate may be impressive, but aren't necessarily appropriate in the worlds of science or medicine. Also be aware that some of the advice offered by wizards and some of the ready-made templates are truly dreadful, contravening all kinds of good taste statutes and aesthetics regulations. Don't rely blindly on wizards or other forms of instant magic. Instead, use the concepts outlined in this book to design effective overheads and slides.

Law of Contrasts

If you have ever flipped through the pages of a fashion magazine, you might have noticed that the only ads that stand out are the black-and-white ones. Because color ads are the norm, advertisers try to differentiate their ads among the kaleidoscope of competing color ads by using the maximum contrast that black and white offer. The same phenomenon is at work in Woody Allen's black-and-white movies, which stand out among the crowd of Technicolor films.

This apparent inversion of the principle outlined above is important: It isn't color per se that is effective but contrast. The human eye and mind respond to contrasts. If everything on a slide or overhead is bright red, everything looks the same. And in a monochrome, nothing stands out, not even your key points.

Formal theories of color and contrast got their start in 1839, when the chemist Michel-Eugene Chevreul published his book *On the Harmony and Contrast of Colors*. Director of a dyeing works, he had noticed that colors' brightness depended not only upon the strength of the dye but also on which colors were placed next to each other. Depending on the color next door, some colors appeared to gain intensity, others to lose it. Chevreul's observations eventually produced the modern-day color-wheel familiar to anyone who has strayed into a home-decorating store.

Spotte Thee Mistaks

Take a good look at the illustration entitled "Spot the Mistaks." How many mistakes can you find?

The illustration's spelling and grammar mistakes are the most obvious, but there are many more. The justification is inconsistent, for example. So is the use of typefaces. Just as too many cooks spoil the broth, using too many typefaces will spoil the overhead or slide. If you use too many typefaces or colors, you get nothing but noise where you want a clear signal. Also remember that any device used for emphasis, such as capitals, typefaces, or color, works because of contrast. Overusing any of these devices negates their power. If you boldface all of your text, for example, you give the viewer's eyes nothing to use for comparison.

Illustration 11.

SPOT THE MISTAKS

i MUST EMPHASIZE THAT THIS POINT REALLY IS ***IMPORTANT*** *BUT SOME OF* ***THE*** *OTHER POINTS ARE NOT REALLY IMPORTANT AT ALL*

ALWAYS PREPARE YOUR AUDIO VISUAL AIDS A LONG TIME IN ADVANCE OF THE ACTUAL PRESENTATION

IN ORDER TO PREPARE YOUR SLIDES PROPERLY, ESPECIALLY IF YOU DO NOT HAVE A COMPUTER GRAPHICS PROGRAM YOU SHOULD DRAW A RECTANGLE IN THE **RATIO OF 2:3** WHICH IS THE RATIO OF A 35MM SLIDE

ALWAYS HAVE A VERY NICE BACKGROUND ON SLIDES, ESPECIALLY IF YOU HAVE ALL THOSE NICE PATTERNS AND COLORS TO CHOOSE FROM IN YOUR COMPUTER GRAPHICS PROGRAM. IF YOU DON'T HAVE A COMPUTER THEN USE A CLR. CRYN. HB4A

POSTERS ON THE OTHER HAND ARE A DIFFERENT MATTER ALTOGETHER WE SHALL DEAL WITH THEM AT ANOTHER TIME

Electronic Presentations

Instead of making overheads or slides, you might want to use direct projection of computer graphics. Microsoft's PowerPoint, Harvard Graphics, and many other software packages designed for presentations can help you go this route. Before you do, consider the advantages and disadvantages outlined in the table below.

Electronic Presentations

Advantages	Disadvantages
Ability to make additions, deletions, or corrections in seconds	Need for access to a compatible projector system
Ability to include animation and other special effects	Poor quality of some projection systems
Ability to carry your presentation around on a disk, transmit it electronically, post it on a network, and so on	Need for a laptop computer or access to a compatible computer
Low cost per slide	Risk of technical failure

Handouts

Some presenters insist on putting up slides with columns and columns of data in tiny print. They then expect the audience to not only be able to see the data but also to perform visual and mental gymnastics in order to give the numbers meaning. Use handouts instead.

When you're creating your handouts, print scaled-down copies of key slides or overheads on the lefthand side of the page. Leave the right-hand side blank for notes. Be sure to leave enough room for note-taking by limiting the number of slides or overheads per page.

Providing handouts allows your audience to concentrate on your message rather than scramble to write down what you're saying. It also gives your audience a hard copy of your ideas, which improves their understanding and retention of your message.

Summary

This section began by explaining the powerful impact that visual aids can have on your audience's attention levels, perceptions of value, and willingness to be persuaded. The section then described ways of using mental maps, key points, and visual metaphors to help you make your case. A series of illustrations provided hands-on experience in improving graphs using design principles outlined in the section. The section concluded with a quick look at electronic presentations and handouts. You can use the strategies outlined in this section for oral presentations or poster sessions—the topic of the next part.

■

Part Four: Poster Presentations

Part Four: Poster Presentations

Part Four looks at a very specialized form of communication: the scientific poster. The section begins by describing the essential differences between oral presentations and poster presentations and the challenges of the latter—including a strict word budget of no more than 500 words. The section then dissects the anatomy of a poster. You'll learn how to analyze your customers; ensure clear, concise communication; and attract browsers to your poster. A section on design principles covers color, emphasis, and other essential elements. The section concludes with a discussion of the often overlooked poster handout and first-aid kit. Even if you never give a poster presentation, the multipurpose tips suggested here make the section essential reading.

The Poster

Andy Warhol once said that in the future everyone would be famous for 15 minutes. He was wrong: It is really 15 seconds. This is the era of the Internet, the 24-hour news channel, and the 15-second sound-bite. The information overload is even more marked in science, medicine, and technology. As the glut of information increases exponentially, there just isn't enough time for everyone to give oral presentations at meetings. As a result, poster presentations have become a permanent feature of today's scientific, medical, and technological meetings.

There are striking differences between a poster presentation and an oral presentation. The most important one? You don't run the show, and you aren't the only show in town. You are like a shopkeeper in a mall, standing by your wares as customers browse. That makes selling much more difficult. Posters do have some advantages over oral presentations. They attract a wide range of people in a less formal milieu than most oral presentations. In addition, people actively seek you and your poster out to get information. They bring tremendous mental energy with them. That makes it a lot easier to transmit your ideas to them.

But first, your poster must capture people's attention. Like movie stars, poster presentations depend on good looks. Unless you and your poster look good, you will be ignored and no one will know about you or your work.

Unfortunately, you have no control over where the meeting organizers will place your poster. You could get lucky and receive a brightly lit location everyone sees the minute they enter the exhibition area. Or you could be consigned to a dimly lit sideroom in the basement of the annex of the spillover building. While you can't control the location of your poster, you can control its design. This part of the book will show you how to make sure you and your poster stand out.

Read the Instructions Carefully

To get the most out of this section, you'll need a watch with a second hand. What follows is a test of your cognitive skills. If you're highly skilled, you'll be able to complete the test accurately in less than one minute, 45 seconds. Average individuals will take up to four minutes to complete the test; the cognitively challenged up to 10. Work as accurately and as quickly as you can. Get a pencil and paper. Clear your desk. Close the door. Start the clock!

1. Working as fast as you possibly can, read through all of the instructions carefully.
2. Fold a piece of paper in half lengthwise. Use the lefthand side to write down the answers to the questions that follow.

3. Draw a square and then draw a circle inside it. Working clockwise and beginning with "A" at the top, label each of the points where the circle touches the square "A," "B," "C," and "D."
4. On the righthand side of the paper, write down the two numbers that come next in this series: 1, 4, 9, 16, 25, 36....
5. Add the two numbers together and divide by two.
6. Check your watch and note on the lefthand side of the paper how long it has taken you to get this far.
7. Write down the fourth, seventh, eleventh, fifteenth, eighteenth, and twentieth letters of the alphabet.
8. Ignore all of the instructions above. Sign your name and stop the clock.

Hands up, all those who didn't read the instructions properly! It clearly said that you had to *read* all instructions carefully. Most meeting organizers send detailed instructions to poster exhibitors. Unfortunately, far too many people ignore or partially ignore those instructions.

Well-organized conference organizers will also send you a checklist for your presentation. However, you often have to wade through a sea of muddled information before you can find it. Create your own checklist. Here are some key points to include:

- Date, day, and time
- Building, floor, and room
- Poster number
- Duration of your poster session
- Dimensions of the board on which you'll hang your poster
- Composition of that board and methods for affixing your poster to it
- Provision of thumbtacks, pins, or other supplies
- Regulations regarding poster titles, presenter titles and affiliations, and maximum number of presenter names allowed on the poster
- Regulations regarding typeface, type style, title size, headings, and text
- Name and pager or phone number of the poster session organizer, in case you have any last-minute problems affixing your poster

Printish vs. English

Well-meaning conference organizers often send instructions regarding the text of your poster. Unfortunately, these instructions are often full of printers' jargon, or Printish. Here's a brief Printish English dictionary:

Printish	English
Case	Capitals are uppercase; non-capitals are lowercase
Typeface	Size or shape of letters
Pitch	Number of characters printed per linear inch
Point	Unit of measure commonly used to indicate font sizes; one point equals 1/72"
Style	Different styles within a given typeface, including bolding, italicizing, and underlining

Anatomy of a Poster

Don't rush out to your local printer or media department or spend hours producing a poster yourself until you have gone through all of the processes described below. A well-designed poster has five key sections: abstract, introduction, methods, results, and conclusion.

Abstract

Although I'm listing the abstract first, write it last. A concise précis of your poster, the abstract should contain your presentation's key points and take-home message.

Introduction

Set your topic in context by writing a concise introduction offering historical perspective. In the great jigsaw of knowledge, where does this particular piece fit? Once you've answered that question, you should clearly state the objectives of the work your poster describes. Use bullets to make your points clearly.

Methods

Use your methods section to show the logic of the model or experiment you used. If you are using mutant Siberian ferrets as a model for Wilkinson's disease, for example, you should explain why these inbred ferrets are a particularly good modeling system for this complex condition. Far too many presenters fail to give adequate explanations about why they're doing what they're doing. You should also briefly note any limitations of the model or experimental system you used. Use bulleted points whenever possible.

Results

Your customers may have already browsed a dozen posters before they get to yours. Don't expect them to digest hundreds of pieces of information and mentally transform them into a meaningful conclusion. Do the work for them in your results section. Instead of showing endless tables of raw data, show aggregate data. Show trends or changes graphically. Use statistics, diagrams, or graphs whenever possible. In short, summarize, clarify, and simplify. If people are really interested in your detailed data, give them a separate handout sheet from your supply in a holder beneath your poster.

Conclusion

Your conclusion section should consist of bulleted points. Answer the question, What is the bottom line? Explain what your findings mean and how they will affect the world. You may also want to give a brief statement regarding the project's ongoing or planned development. If your work is really taking off, use the poster as an opportunity to transmit your excitement about it. The next browser at your poster could be the vice-president of rodent development at Bucktooth, Inc., looking for his next big investment. If you present well, it could be *your* patented Intercontinental Ballistic Ferret that is the next big hit in the field.

The Fine Art of Writing

Word Budget

Many people go wild when they put together a poster. Instead of headlines and key points, they write a novel. They have an eight by five foot area to fill, and they intend to fill it. Avoid this temptation. A poster needs to be as concise as possible—500 words at most. "No way!" you gasp. "My abstract alone is longer than that." If that's true, rewrite your abstract. It is too long.

Why only 500 words? Your customers have probably browsed a dozen posters before they get to yours. If they approach and see nothing but line after line of text with little or no white space, you will lose them. To sell your work and yourself, you must be customer-oriented. That means your product must be accessible. Get rid of the non-essential ink and show only the key facts and findings.

How can you boil your message down to 500 words? Build yourself a framework based on Sir Austen's questions as described below. Answer the questions, then rewrite, rewrite, rewrite. Apply the rules of simplification and clarification at every round.

If you can't say everything in 500 words, you are trying to say too much. Think about the basic message you are trying to get across. Are you trying to transmit more than one basic message? Are you trying to tell the world everything you have ever done all in one gloriously muddled poster? If so, you must start again at the very beginning. Go through the brainstorming process, then separate the extraneous and the essential. Ask yourself what a nonspecialist wants to know about your work.

Using pictures can also help you tell a story with only 500 words. If a picture paints a thousand words, then a set of carefully designed graphs, diagrams, or illustrations can say as much as any novel. In fact, a single, well-designed graph can often say it all. It is a visual summary of your data, an easily digestible picture of your findings.

Sir Austen's Quintessential Questions

Sir Austen Bradford Hill was a rare phenomenon: a clear thinker. He devised four simple questions that you should ask yourself as you try to explain your work. Use these questions as a framework you can build upon:

- **Why did you do it?** That may seem like an obvious question. But all too often, presenters forget to explain why. Instead, they rush right into presenting their results.

- **What did you do?** Again, a deceptively simple question. If it's so obvious, though, why do so many presenters fail to answer it in clear, concise words or diagrams? How exactly did you conduct your study? What were the key stages involved? How did you measure the key parameters?

- **What answer did you get?** Do not bombard your audience with every piece of raw data that you generated. Instead, summarize what you found out. Aggregate your data. Perform statistical tests when appropriate and use graphs or diagrams to report your results. But beware! It's amazing how many presenters merrily perform regression analyses but never bother to test whether their data are normally distributed. Many other crimes against statistics go unpunished and undetected even after an alleged peer review and publication in a high-quality journal. Whenever possible, get a statistician to advise you on appropriate statistical methods.

- **What does it mean?** This is the trickiest part of all. What is the significance of your finding? Do not fall down at the last fence. Give meaning to all the wonderful work you have done. Follow Sir Austen's framework, and I guarantee you'll give a more effective presentation.

Clear and Concise

It's surprising how many people honestly believe that they can write well. They churn out memos, lecture notes, letters, papers, and abstracts by the hundreds without ever giving much thought to their writing style.

This is unfortunate, because they are hampering their ability to communicate. Not everyone is going to be a Tolstoy or a Dickens. However, you can ensure clear, concise communication by following a few simple rules:

- Choose your words carefully. Use one word instead of two or more, whenever possible.

- Use nouns and verbs instead of adjectives and adverbs.

- Always choose the concrete rather than the abstract, providing specific examples rather than talking in abstractions. For example, you should rewrite sentences like "One cannot exaggerate the importance of measuring the core temperature of one's ferrets preceding, during, and following the experimental scenarios described herein" as "We recorded each ferret's core temperature."

- Avoid tautologies, such as free gift (gifts are free by definition), track record (record will do), or live audience. (You hope they will still be alive at the end of your presentation!)

- Keep sentences short. Remember, you are working with a total word budget of 500 words max. You want your message to be succinct and easily digestible in the shortest possible time. You do not want people dying of old age halfway through panel 9,994 of volume eight of your poster.

- Use single words for subheads or subtitles. This helps keep you within your 500-word budget and forces you to focus on the exact purpose and meaning of each of your poster's sections. If you can't describe the section's function in one word, such as "Methods" or "Results," it probably has more than one function or contains extraneous material that is not really essential to that section's message. Divide the section in two or delete extraneous material.

Standing Out in a Crowd

How do you attract customers to your poster? Try these strategies.

Title

Titles are important. Just ask any king, duke, lord, knight, or baron. Like the headline of a newspaper article, your poster's title is a hook that snares your audience's attention. Keep in mind that your poster's title will be used in all the "advertising" that precedes your presentation.

Stimulate people's interest and whet their appetites by choosing a provocative, challenging, or witty title. Your title is your herald, a harbinger of your message.

Keep your title short, no more than eight words. Short titles are more memorable and pack more psychological punch than long ones. Think about book and movie titles: *Speed, Gone with the Wind, Sense and Sensibility*. Avoid pompous titles, such as "Toward an Understanding of...." This is not the nineteenth century. Also avoid such titles as "Blah Blah Blah Past, Present, and Future."

Abstract

Write a snappy abstract. Remember, most meeting participants look through poster titles and abstracts when registration materials are distributed. Because participants know there will be far more information at the meeting than they can possibly absorb in the time available, they decide in advance which presentations look inviting and which look dull. That's why your title and abstract are important advertising tools.

Meet and Greet

You're not done once you've thought up a catchy title and a thought-provoking abstract. Once you're in the poster exhibition hall, you and your poster must continue standing out. Dress appropriately for your poster presentation. Then smile and greet every browser just like salespeople do in the better-run stores. A welcoming smile and brief hello can break down barriers and encourage nervous or hesitant browsers to check out your poster. Don't turn people off by looking somber, dour, or bored as you stand by your poster. If you look bored, your customers will conclude from your body language that your work isn't worth checking out. Stay by your poster as long as possible during your allotted time.

Know Your Customers

Successful store owners know their customers. What do you know about yours?

The most startling fact is that 95 percent of the people browsing your poster are not specialists in your field. Some presenters think that's just too bad. Those people should have gone to the free demonstration of the new Pangalactic Hyperwhiz Omnianalyzer instead of bothering me with questions about my poster, they grumble. These presenters assume that only the five percent of browsers who are experts in the field can help them with ideas or share findings.

Wrong! Do not ignore browsers just because they aren't experts in your field. Some of the greatest scientific breakthroughs have come about when concepts developed in one field were applied to an entirely different field. It's in your interest to attract the 95 percent as well as the five percent. Make your poster clear, simple, and attractive as well as exciting, provocative, and intriguing.

To do so, keep abbreviations, jargon, and acronyms to an absolute minimum. If they are unavoidable, be sure to define them the first time they occur. If you follow this book's suggestions, your presentations should be accessible to 100 percent of the people who attend them no matter what their expertise. Remember, the real goal of a presentation is to communicate your ideas and get some good ones back again.

Designing Your Poster

Long before you start working on your poster's layout, think about its physical design. How you plan to get your poster to the meeting, for instance, can affect your design choices.

Logistics

Will you and your poster travel to the meeting by car? Motorcycle? Train? Plane? By being fired from a cannon? The disadvantage of a large single-sheet poster lies in its aversion to being transported easily. You must roll it up in an unwieldy cardboard carrying tube, which is especially hard to transport by airplane. You'll probably have to check it as "special luggage" along with the skis, cats, dogs, and ferrets.

This could prove disastrous. Airline baggage handlers are not renowned for their care when hurling your luggage around airports. Airlines also lose luggage on occasion, usually on very important occasions! I always carry my presentations with me, whether they're slides, overheads, or posters. If it's a poster, I design it in equally sized panels and pack them as carry-on luggage. That way they never leave my side. In addition, I always carry a suit, tie, shirt, dress shoes, change of underwear, and shaving kit as part of my carry-on luggage. No matter what happens to the rest of my luggage, I can still give my presentation and even be dressed for the part. Always assume the worst and plan accordingly.

Hunting and Gathering

Collect your poster's components. Do not glue, nail, or weld anything down yet! Components might include your data (in graph or table form wherever possible), charts, diagrams, photographs, and text. If actual components aren't available yet, use scale representations.

The next step is to arrange the components. There are several ways to do this. Some people like to plan their posters by working at full scale. You could lay out the components on a large table that you have marked with the dimensions of your allotted space, for instance. (Double-check the dimensions in the instructions given by the meeting organizers.)

Other people prefer to draw a scale diagram of the poster onto graph paper and then cut out scale representations of each of their components. This is similar to redesigning a room in your home. Instead of moving actual furniture around, you use a scale drawing and paper representations of your furniture to redesign your home's layout until you get it just right. Computers can simplify the process even further. Use a graphics program to model your poster and its components.

Whatever method you use, focus on fit. Will the components fit in the space available with room for white space and perhaps mounting boards as well?

Do It Yourself

Some presenters have highly skilled media departments that can help them produce their posters. Other presenters have limited resources and produce the entire poster themselves. That's all right. You don't need a media department to produce a first-rate poster. You can do the whole thing yourself.

Having access to a computer helps. A profusion of software allows anyone to produce a high-quality poster. Specialized graphics programs, such as Harvard Graphics or Corel Draw, are not essential. Most of the leading word-processing packages can produce poster-size text. If you don't have access to one of them, pick up one of the cheaper programs. Key Publisher, for example, is a value-packed bargain. Numerous freeware and shareware programs are available on disk or via the Internet.

Having a high-quality color printer also helps. If you don't have a printer capable of printing large fonts, take your final text and graphics to a print shop for printing.

If you don't have access to a computer, you must rely on a typewriter. At least try to beg or borrow an electric typewriter. Type what you need, then enlarge the text by using the best-quality photocopier you can find. Select the thickest typeface available, because thin typefaces don't look good after enlargement by photocopying. And choose your thickest paper, because thin paper distorts the area around each key impact site. Use a new ribbon when you're preparing the final version and triple-space your text. Be sure to clean the photocopier's glass before you use it. You want to enlarge your text, not the detritus left over from previous users.

Readability

Test the readability of your final text. Is it readable from a meter away? If not, it's too small for a poster. Remember, you want your poster to be as user-friendly as possible. Readers should not have to press their noses against your poster in order to read it, nor should they need their portable electron microscope.

Use no more than five to six words per line. This may seem like cruel and unusual punishment, but a poster is not supposed to be a wallpaper essay describing your work. It is information that has to be read and understood in a short time by tired people who must not only stand while reading, but must also mentally block out the hubbub going on around them. Obviously these are not ideal conditions for transmitting your ideas effectively. When reading posters, our eyes are less able to scan accurately from line to line than when we are reading text in books or journals. They become easily confused and either jump back to the beginning of the same line or skip a line altogether. Keeping sentences short reduces the demands on your reader.

Short sentences also mean more white space. Too much text turns readers off immediately. White space gives the reader cues and helps divide the text into logical, mentally digestible, bitesize pieces. Do your visitors and yourself a favor: Make sure *all* text is readable from one meter, not just the headings and title. Remember, your total word budget is 500 words. That leaves plenty of room for large type!

Color

Used carefully, color can help make a good poster outstanding. Even the do-it-yourselfer can use color. If you don't have access to a color printer, deliver your disk or send your file via modem to a print shop for color printing.

Color can provide mental cues that enhance your poster's clarity and accessibility, thereby demanding less mental effort on the part of your readers. You might use color to emphasize themes, for example, or use it to show patterns in your data. Try using color to emphasize your introduction and conclusion. Simply use mounting boards that are a different color than the rest of your panels.

It's important to be consistent in your use of color. If you use red to identify your experimental group in one chart, for example, use red to identify that same group in all your other charts and graphs.

Capital Punishment

Pay careful attention to instructions about typeface, point size, and style. You may be absolutely crazy about Ultra Shadow or Monotype Corsiva, but they're not suited to a scientific poster presentation. The most commonly encountered typefaces—Times Roman, Helvetica, and Elite—are also the clearest.

Don't get carried away by your computer's power to generate text or graphics. Keep the number of different typefaces to an absolute minimum. If possible, use just one typeface throughout and use larger type, boldfacing, italicizing, or underlining for emphasis. Like too many words, too many typefaces can overwhelm and confuse the reader.

Don't punish your audience by making them read sentences that are made entirely of uppercase words. Uppercase text is difficult to read. It tires the eye and hurts the brain. What have your viewers done to deserve this cruel (and all-too-common) punishment? Use capital letters only at the beginning of sentences or in widely accepted or clearly defined acronyms and abbreviations.

Logical Flow

English, the international language of science, is read from left to right and from top to bottom. Your poster's layout should follow these conventions. Don't make your customers' brains hurt as they try to unravel your poster's logical flow.

Your poster will be a series of panels. Whenever possible, make the panels of equal size by mounting the components onto stiff cardboard or board. Using equally sized panels makes the process of designing your poster far easier. It also allows for a more aesthetically pleasing, less cluttered look.

Once you've decided on the final layout, number each panel in sequence on its top lefthand side. These numbers will give your readers cues about your poster's logical flow. Remember to leave room at the very bottom of the righthand poster to attach a holder for your handouts. (The holder can hang from your poster board's bottom edge.)

Once you have draft versions of all of your components ready, lay them out on a table and ask a colleague to read your poster. Ideally the colleague should not be an expert in your field. Using a nonexpert at this stage is a good way of ensuring a highly accessible, easily understood finished product.

Ask your colleague to read right through, without comment or interruption. This will allow him or her to give you a feel for your poster's overall readability. Then ask for specific criticisms: Is the abstract comprehensive? Is the description of your methods clear? Are the data presented clearly? Does the significance of your results come across? Is the take-home message clear? Also ask for feedback on the poster's clarity and logical flow: Does the poster work?

Make the appropriate changes, perhaps deleting a table here and inserting a graph there, defining a term here and deleting jargon there. Then go through the process again, preferably with a different colleague who hasn't seen your poster already.

Revise your materials over and over again. Keep only those components that add value to the poster by increasing its clarity and accessibility. Reject the rest. No matter how beautiful the three-dimensional Technicolor graph you made for last year's presentation is, don't be tempted to use it just because you have it. Use materials only if they fit with the logical flow of this poster presentation.

Essentially you are using the Goldilocks approach to design: Too much information, and your poster will lose clarity. Too little, and it won't make sense. Like Goldilocks and her porridge, you're trying to find the amount that's just right.

Double-checking

Remember your checklist? Now's the time to get it out again. You need to be absolutely sure that you have followed the conference organizer's instructions and avoided other mistakes. Recheck the typeface and size, the title size and format, the authors' names and affiliations, and the point size of headings, text, and acknowledgments. Check for typos and grammatical errors. Check that you have actually said what you meant to say!

Ask a colleague to look over your poster again. New eyes can often spot mistakes that familiar eyes fail to see. This may all seem like overkill and a lot of effort for a poster that will, after all, only be on show for a few hours at most. Perhaps. But you need to think of this poster as a huge, glaring, public billboard of you. If it is muddled, dull, or sloppily done, you look bad. If it looks crisp, professional, well-organized, clear, and easily understandable, you must be, too!

Mount Up!

Whenever possible, mount your poster's components on stiff cardboard or board to form panels. In addition to protecting them, it helps the components stand out better and gives your poster a more coherent, polished look. Mounting all your components on the same size boards makes arranging the poster easier, too.

The size of your panels will depend upon how you are preparing your poster. If it is a do-it-yourself poster, you will probably be using 8½ x 11 inch paper for your text and graphics. Make sure the mount is at least two inches greater than the paper's dimensions. This gives the mount enough "presence" for the eye to notice. Any less, and the mount can't frame its contents as effectively.

Choose a mount whose color complements the dominant colors of your components. Avoid using white or bright colors for your mounts. These are harsh on the eye and detract from your poster's user-friendliness.

Once you've decided on the exact order of your panels, add numbers. Use Letraset or similar rub-on products to number the upper left-hand corner of each mount.

Handouts

"Handouts?" you say incredulously. You may rightly ask. Very few presenters ever bother to produce handouts. That's too bad. Even if they produce a wonderful poster that generates a lot of interest, that interest is often ephemeral. The average visitor does not have the time to write your ideas down.

Make things easy for them. Produce a one-page handout, making it double-sided if necessary. Visitors love getting a take-home summary of the work being presented. They don't have to take notes, and they know they can easily contact you in the future. You can also make detailed data available on your handouts. You can even include last-minute updates. Handouts are polite. And they ensure that people remember you and your work.

Always include your name, address, phone and fax numbers (including area and country codes), and e-mail address on the handout itself. In addition, staple your business card to the upper righthand side of each handout. This may seem like overkill, but many people like to keep business cards in their files for future reference. They will separate your card from your handout. Sometime in the future they may remember your work but not your name, and there it is on both the handout and in their business card file. Making it easy for others to contact you is an essential element of effective networking.

Place your handouts in the holder you attached to the bottom righthand side of your poster when you designed it. Bring a large number of copies. It is better to over-estimate demand and be left with extras than to underestimate and have nothing to offer the person who was just about to offer you a job or a grant for $100,000!

First Aid

Never assume that your poster will arrive in pristine condition, no matter how carefully you pack it. Pack an emergency "first-aid" kit with your poster and bring it to the poster session. Your kit should include:

- A box of thumb tacks—even if the organizers assure you they will provide tacks
- A retractable glue stick
- Adhesive tape on a dispenser
- A pair of scissors
- A black pen
- A notepad
- Correction fluid
- Spare business cards
- Breath mints
- Small package of Kleenex

Use the scissors, glue, and tape for any last-minute repairs you need to make. Use the pen and paper for taking notes. Be sure your pen is black: No matter how carefully you checked your poster, you won't notice any errors until you're actually exhibiting your poster. The black pen and white correction fluid may be your salvation.

The spare business cards are there just in case your poster is more popular than you anticipated. Remember, there is nothing more unprofessional than scavenging around for a pen and paper to jot down your address for one person after another. Always err on the side of caution. Business cards are small and not a burden to carry with you. Better to take too many cards than too few.

The Kleenex can come in handy if you get an unexpected case of sniffles, thanks to an inconvenient cold, dust, allergies, perfumes, or anything else. You do not want to be dripping mucus while trying to explain your work. The breath mints are insurance. Remember, you may be nervous about presenting your poster or getting it up before the madding crowd is unleashed upon you. You may have had garlic for lunch. Whatever the cause of your halitosis, be sure to have some breath mints with you. You will be talking to people one-on-one at very close range. You want them to remember your work, not your bad breath. Play it safe even if you're convinced you never get bad breath.

Go Browsing Yourself

The most annoying thing about giving a poster presentation is that others who are working on topics related to yours are showing their posters at exactly the same time you are showing yours. If you stay at your poster throughout the entire period you've been allotted, you will never get to meet people who might influence your work or your career profoundly. If other members of your team are present, you can take turns "manning" or "womaning" the poster. If you're the only team member around, there's no easy answer to the dilemma. Once the crowd thins out a bit, you could affix a "Back in 10 minutes" sign to the top righthand corner of your poster. People who stop by while you're gone can come back later or can simply take a handout and business card from the holder attached to your poster.

Don't use a sign that's handwritten in crayon. Prepare this sign at the same time you print the rest of your poster. Remember, everything you do is an advertisement for you. Always present well, even when you are not there in person!

When you set out to visit the other posters, take your business cards, a pen, and paper with you. If a particular poster or presenter interests you, explain why you can't have a long discussion and jot down where they are staying. You can always arrange to meet people later. Exchange business cards, so that if you fail to meet again during the meeting you know how to find each other.

Recycling

Posters have limited use. Despite the amount of time and energy that you put into producing them, they are on show for a few hours at most. You can make better use of a poster by recycling it.

Of course, using the poster at another meeting would be unethical. But posting it in your department, lab, or other work space extends its life-span. Often your colleagues aren't fully aware of exactly what it is you have been doing all these years. This way they can see. Leave the poster up for a while to give everyone, including visitors and your boss, the opportunity to see your work. This is good advertising for you.

Summary

The book's final section dealt with a specialized form of communication: the scientific poster. The section began by offering tips for boiling your message down to the 500 words recommended for posters. The section then dissected the poster's anatomy, including the title, abstract, introduction, methods, results, and conclusion. A section on design principles covered color, emphasis, and other essential elements. The section concluded with a discussion of poster handouts and first-aid kits.

■